MATHEMATICS
FOR THE
LAYMAN

MATHEMATICS

FOR THE LAYMAN

BY

T. H. WARD HILL M.A.
SOMETIME SCHOLAR OF QUEEN'S COLLEGE OXFORD

ROWMAN & LITTLEFIELD PUBLISHERS, INC.
Lanham ▪ *Chicago* ▪ *New York* ▪ *Toronto* ▪ *Plymouth, UK*

Published by Rowman & Littlefield Publishers, Inc.
4501 Forbes Boulevard, Suite 200, Lanham, Maryland 20706
www.rowman.com

10 Thornbury Road, Plymouth PL6 7PP, United Kingdom

Originally published by Philosophical Library
Copyright © 1958 by PHILOSOPHICAL LIBRARY, INC.
First Rowman & Littlefield paperback edition 2014

British Library Cataloguing in Publication Information Available

Library of Congress Cataloging-in-Publication Data Available

ISBN: 978-1-4422-3418-5 (pbk. : alk paper)
ISBN: 978-1-4422-3419-2 (electronic)

♾™ The paper used in this publication meets the minimum requirements of
American National Standard for Information Sciences—Permanence of Paper
for Printed Library Materials, ANSI/NISO Z39.48-1992.

Printed in the United States of America

PREFACE

FIGURES simply mean nothing to me; I never had any head for mathematics." All of us have heard, or made, this sort of remark; some people almost seem to take pride in saying it. And behind it there lies the suggestion that mathematicians are a race apart, people who, if they do not actually deal in black magic, nevertheless come very near it.

Now is this really so? The person who is "no good at figures" can often check a bill with greater speed and accuracy than the professional mathematician. And the woman who "never could see the point of learning geometry" will often be most proficient in arranging her furniture, in thinking out new schemes of decoration for the home, or in appreciating beauty.

Are we all so deficient in mathematical knowledge as we often think we are? Is it all so unnecessary, is it such a terrible discipline? No; all of us use mathematics in some form or other every day, in our work—and often in our play. There are not many people who can resist the puzzles which are such a feature of our newspapers; and why do such hot arguments arise over the correct answers to 'teasers'?

The answer is not far to seek. Mathematics is something fundamental in our lives. For mathematics is a tool—a set of tools, if you like. We learn to use those tools so as to bring order into our lives. It is not necessary for us to know how those tools have been forged, though this may often be a help to us. We ride a bicycle or drive a car; we do not have to know the principles of mechanics, or the theory of the internal-combustion engine, to do this.

So we have to learn how to use the tools of mathematics. And as we learn we shall perhaps be given a glimpse of the unity that is mathematics, some idea of the purpose behind it all. We may see how, from modest beginnings, a mighty edifice has been built up; how, by what is known as generalization, groups of apparently isolated facts can be shown to be particular cases of one master-fact or theory; how, by discarding irrelevant details, we are led to the essentials of a problem.

No one will expect to find a book about mathematics which can be read quite like a novel or a history book. But we shall try to approach the subject, from time to time, from a novel angle. We shall see, for instance, how the development of mathematics has been retarded because for hundreds of years there was no convenient system of shorthand as we now have in algebra; and we shall see how a lucky inspiration has sometimes opened up new fields of knowledge, as when Descartes, three hundred years ago, had the vision which led eventually to the graphs with which we are becoming increasingly familiar in our daily lives.

Much there may be that has been done before. But in the background there must always be pen or pencil and paper. "Practice makes perfect," so some practice there must be. This book will not be flooded with exercises, as text-books necessarily are. But those that are given will serve a double purpose. They will help the reader to check if he is understanding what he is reading; and their successful solution may give the reader just that encouragement necessary for him to go ahead—and perhaps find that mathematics is not such a closed book to him as he may have thought.

T. H. W. H.

ACKNOWLEDGMENTS

THE author wishes to thank Burroughs Adding Machine, Ltd, for permission to reproduce the illustrations at pages 15, 16, 18, 21, 67, 86, and 290, and for kindly reading the proofs of the section on calculating machines.

The extract from *Introduction to Mathematics*, by Professor A. N. Whitehead (Home University Library), is reproduced by permission of the Oxford University Press.

The tables of logarithms, etc., at the end of this book are reproduced by permission of Mr J. T. Brown, M.A., B.Sc., and Mr A. Martin, M.A., B.Sc.

The author's sincere thanks are also due to his publishers, without whose very ready co-operation the labour of producing this book would have been very considerably increased; and to his former colleague, Mr D. W. Pye, M.A., for his valuable help in reading the proofs.

CONTENTS

PART I

NUMBERS—NUMBERS—NUMBERS

MATHEMATICS
FOR THE
LAYMAN

PART I

NUMBERS—NUMBERS—NUMBERS

CHAPTER I

IN THE BEGINNING

THIS is not a history of mathematics, nor is it intended to observe a strictly chronological order. The origin and development of each branch of the subject will be outlined in the appropriate place. Before we come to the particular, however, something must be said about the early history of mathematics as a whole.

We know that some sort of commercial arithmetic was used in Babylonia over 2000 years before the birth of Christ. In that country records of business transactions corresponding to our invoices and bills and ledgers were scratched on clay tablets, and these were baked to make them hard. Many of these tablets have been deciphered, and some of them give almost complete records of the business done by great commercial families. Other tablets which have been found concern property belonging to the temples.

SUMERIAN CLAY TABLET

In the centre is the number 6; below this, represented by two circles and four crescents, the number 24.

They were great business men in those days; laws had to be made to regulate commercial dealings.

We have records of mathematics from Egypt, too, long before the Christian era. They were either carved on stone or written on papyrus. Thin strips were cut

15

lengthwise from the stem of the papyrus, which is a plant resembling a reed. Rows of strips were first laid out side by side; then a second layer was added, the strips being placed across the first. A weight was placed on top, and the strips were allowed to dry. During this process the sap of the plant glued the sections together,

AN EGYPTIAN SCRIBE WORKING ON PAPYRUS

and the finished papyrus looked rather like a rough brown paper. Writing was done with a reed pen.

What is to-day the most famous of these mathematical works was copied in about 1650 B.C. by Ahmes, an Egyptian priest, from an older work. It was brought to this country about a hundred years ago by Henry Rhind, and it may now be seen in the British Museum. An amusing feature of the papyrus is that corrections in red

16

ink appear on it, which were probably made by the hand of a teacher.

The Ahmes, or Rhind, papyrus has the curious title "Directions for knowing all dark things, all mysteries, all knowledge." It consists chiefly of arithmetic and practical geometry, though there are also easy problems, which nowadays we should solve by algebra, such as:

A quantity, its whole, its seventh, it makes 19.

Ahmes deals with fractions, and he gave rules (often inaccurate) for finding the areas of the commoner surface-shapes.

His geometry was essentially practical; it was concerned with land division and building. The surveyors of those days did remarkably well when we consider the crudity of their measuring instruments.

Long before this man had sought to decorate and beautify his surroundings and the objects which he used in his daily life. Influenced, no doubt, by the many beautiful forms of nature—trees, a spider's web, the frost, for example—he endeavoured, slowly and crudely at first, to experiment with ornament and shape. And so it was that in the heyday of the civilizations of Babylon and Egypt gorgeous and intricate geometrical designs were freely used in jewellery and ornaments of precious metals and stone and in the pavements and walls of cities that for long after remained hidden. They did not know geometry as we know it, but they knew how to apply geometric design in their search after beauty.

So we see that from the earliest times mathematics has had a twofold influence on the lives of people: the practical and the æsthetic. The great merchants and priests each contributed their share to its early development. In passing, it may be mentioned that it was well that so much of this development was enshrined in the priestly mysteries, or else much of it might have disappeared in the dust, and a start of thousands of years would have been lost.

We shall read later of how mathematical knowledge grew—how, for instance, the Greeks largely changed it from practice to theory. It is an interesting story—how man forged an instrument which, in its growth and perfecting, has made possible the great advances which have led to our modern civilization, and which at the same time has owed much of its growth and perfecting to the demands which our modern civilization has made upon it.

BEFORE HISTORY: THE CAVEMAN CALCULATES WITH
A CHARRED STICK ON A ROCK-FACE

We know the names of many men who helped to develop mathematics; we cannot say this of the men who gave us our number-system. But there is no doubt that in whatever country a system of representing numbers was invented it was originally dependent upon counting with the fingers. And just as writing came long after speech, so the writing down of numbers, or signs to represent numbers, came long after early man had learned to count.

The earliest numbers were notches on a stick, or scratches on a stone, or marks on pottery. These would be imitations of the fingers, and early ways of writing down one, two, three, and four would just be the repetition of these numbers of vertical or horizontal strokes. The word 'digit' for the figures 1 to 9 undoubtedly comes from this source.

Write two strokes thus (=). Do the same thing with three strokes (≡). If you do this quickly, it is quite easy to see that we can trace the development of our present figures 2 and 3 from that source.

An extension of finger numbers was being used by merchants as late as the Middle Ages when they traded with foreigners. The picture on page 22, which is taken from a book written in 1494, shows the way in which fingers were used to show different numbers.

All the countries of the ancient world had their number-symbols. The Babylonians used number-pictures depending for shape mainly on the wedge and the circle. There were two distinct sets of Egyptian numerals; the Greeks had a letter-system and also two systems in which peculiar symbols were used. But in these and other systems, including the Roman, the numbers one, two, and three were depicted by three strokes of varying shape, and there was a special symbol for the number ten.

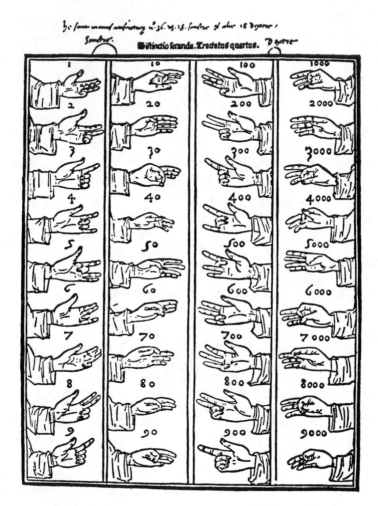

FINGER NOTATION FROM PACIOLI'S "SŪMA" (1494)

The Roman system is more familiar to us; it is still used on the faces of clocks and watches, and old mile-

ROMAN BONE COUNTERS, NOW IN THE BRITISH
MUSEUM, ENGRAVED TO SHOW THE NUMBERS
8 AND 9

stones with the distances carved on them in Roman numerals may still be seen.

Our present number-system is generally called Arabic, but it should properly be called Hindu-Arabic. It prob-

I	II	III	IIII	V	VI	VII	VIII	VIIII
1	2	3	4	5	6	7	8	9

X		XXX	↓	↓X		↓XXX
10		40	50	60		90

C		CCCC	Ð		ÐCCCC
100		400	500		900

⚭	⊖	⟪ʋ⟫	⟨⟩	⟨⟨ʋ⟩⟩
1000	1000	10,000	50,000	100,000

ANCIENT FORMS OF ROMAN NUMERALS AND THEIR PRESENT-
DAY EQUIVALENTS

ably originated as a set of cabalistic (secret) signs invented by Hindu priests. From India these signs were carried to Baghdad, and from there Moorish merchants brought them to Spain, where they are first seen in a book produced in A.D. 976.

Despite its obvious utility, the Hindu-Arabic system was only slowly recognized. About A.D. 1300 we read that bankers in Florence were forbidden to use the numerals, and booksellers were obliged by law to mark their wares "not in cyphers, but in intelligible letters."

Although ten (ten fingers to our hands) is the basis of most number-systems, traces of systems based on three, four, five, six, eight, and twenty are still found among primitive tribes.

We, and also the French, still retain traces of a system based on twenty. 'Three score years and ten' and *quatre-vingt* (four twenties) come immediately to the mind. A peculiarity of numbering in Welsh is that sixteen, seventeen, and nineteen become (in translation) one

INDIAN	
ARABIC (ELEVENTH CENTURY)	
OLD ITALIAN	
ITALIAN (FIFTEENTH CENTURY)	
MODERN ARABIC	

THE NUMBER 1942 WRITTEN IN CIPHERS

$(10+10+10+1+1) \times 60 + 10 + 10 + 1 + 1$

BABYLONIAN (CUNEIFORM)

一 千 九 百 四 十 二

$(1 \times 1000) + (9 \times 100) + (4 \times 10) + 2$

CHINESE

$1000 + \begin{matrix} +100+100+100+100+ \\ +100+100+100+100+ \end{matrix} 100 \begin{matrix} +10+10 \ +1 \\ +10+10 \ +1 \end{matrix}$

EGYPTIAN (HIEROGLYPHIC)

$\alpha \quad \gg \quad \mu \quad \beta'$

$1000 + 900 + 40 + 2$

GREEK

M C M X L I I

$1000 + (1000 - 100) + (50 - 10) + 1 + 1$

ROMAN

**THE SAME NUMBER (1942) IN ARITHMETICAL
CHARACTERS NOT GOVERNED BY POSITION,
WITH MODERN EQUIVALENTS**

and fifteen, two and fifteen, and four and fifteen, and the word for eighteen is *deu-naw* (two nines).

In all the ancient systems much trouble could have been avoided if some one had thought of a symbol for zero. To us, who use 0 without giving it a thought, it must seem strange that it was such a very long time before anybody invented such a symbol.

It is not known exactly when the zero-symbol was invented. It was at least as early as the ninth century A.D., and it may have appeared as early as the second century B.C. But it was not invented to signify nothing, in the sense that you may have nothing in your pocket. It was used as what the Americans very properly call a 'place-holder.' By its introduction man was enabled to extend his concept of numbers indefinitely.

CHAPTER II

NUMBERS AND NUMBER-FORMS

WE are so accustomed to using numbers merely in connexion with calculations—those involving just numbers, and those in which numbers are used to denote multiples of unit values (as in money, weights, and measures)—that it may come as a surprise to some readers that the study of numbers themselves is of great antiquity. The books of the Bible contain all sorts of queer combinations of numbers; they all had some special meaning at the time when they were written, and, indeed, a whole literature has grown up about their interpretation. People wonder, for instance, at the frequent use of the number seven in the early Christian writings—the seven deacons, the seven deadly sins, and so on; and, later on, the seven Sacraments, the seven Orders, the seven joyful mysteries and the seven sorrowful mysteries, to name but a few.

Some people even nowadays prophesy queer things by studying numbers. Who has not read of the calculations made from the dimensions of the Great Pyramids of Egypt? And, let it be whispered, what of the motorist who takes a delight in adding up the numbers on the identification plates of passing cars, and making his own deductions therefrom?

THE NAME 'ARITHMETIC'

We have already seen that the Babylonians and Egyptians used what we should call commercial arithmetic. The Greeks and the Romans added little to their knowledge, but they liked 'playing with numbers.' So keen were the Greek scholars on studying the properties of numbers that they gave this study a special name. They called it $ἀριθμητικη$ (*arithmeetikee*), a word derived from $ἀριθμος$ (*arithmos*), the Greek word for number. The practical, everyday calculations of the merchant they called 'logistic.' It is interesting to note that the current American military term for the science of moving and quartering troops is logistics.

ADAM REISE

The author of *Rechnung nach der lenge auff den Linihen und Feder*, published in Leipzig in 1550.

This division of the subject continued for hundreds of years, right up to the Middle Ages. It was not until the close of the fifteenth century that books were brought out which combined the two.

On the continent of Europe the word arithmetic still carries the old Greek meaning. What we, and the Americans, call arithmetic is described as *Rechnung* in German and *calcul* in French, both words meaning calculation or computation.

Some readers may be interested in the following examples of the way the Greeks (and others since) tried to play with numbers. Not all of it was useless! The more practical-minded may like to pass over the next three sections, which deal with 'perfect' numbers, 'prime' numbers, and 'figurate' numbers.

PERFECT NUMBERS

The Greeks had special names for special numbers.

6, for instance, was called a 'perfect' number; a perfect number is a number which is equal to the sum of all its possible factors, including the factor 1. (2 is called a factor of 6 because 6 can be exactly divided by 2; similarly 3 is a factor of 6.) Perfect numbers are not easy to find; 28 is the next one. Test it for yourself. The next three perfect numbers are 496, 8128, and 33,555,336.

PRIME NUMBERS

Much work was also done on 'prime' numbers—that is, numbers which have no factors other than themselves and 1. Eratosthenes (about 230 B.C.) invented a device called a *sieve* for locating them. His scheme consisted of writing down all the odd numbers (why only the odd numbers?) and crossing off every third one after 3, every fifth one after 5, and so on. The method is laborious, but it works, and no better method has been found.

There is no end to the number of prime numbers. A simple form of the proof of this runs as follows:

Multiply together all the numbers in the 'run' from 1 to (say) 10. Add 1 to the product. The result must be a prime number. This can be done for any 'run' of numbers starting from 1. So we can form as many prime numbers, or primes, as we please.

FIGURATE NUMBERS

Pythagoras (sixth century B.C.) pictured numbers as having orderly designs. There were the *triangular numbers*, 1, 3, 6, 10, . . .

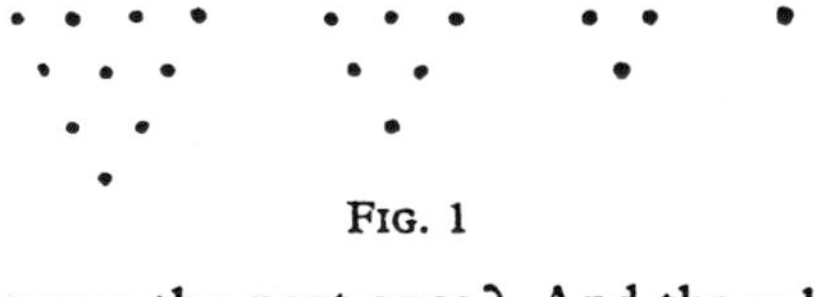

FIG. 1

Can you discover the next ones? And the rule by which each is obtained from the preceding one? This is much more interesting than saying that the numbers 1, 3, 6, 10, . . . are obtained by adding consecutive whole numbers starting from 1. Yet both are aspects of the same truth.

Then there were the *square numbers*:

Fig. 2

Each number (as depicted by the number of dots) is obtained from the preceding one by adding a border shaped like an ∟ reversed—thus ⌐.

Notice that the number of dots to the right of and below each consecutive ⌐ gives us the odd numbers in order. Can you form the next square number in the above manner?

From this we obtain a generalization—that is, a property which holds generally:

The sums of consecutive odd numbers starting from 1 give the square numbers 4, 9, 16, . . .

In the same way the Pythagoreans said that the sum of consecutive even numbers made the *oblong numbers* 6, 12, 20, . . . Do you see how they are found here?

Fig. 3

And can you give the next two oblong numbers?

SYMBOLS AND SYMBOLISM

It is not generally realized how much we owe to the inventors of the symbols (or mathematical shorthand signs), without which mathematics would be much more difficult to understand than it actually is. Where should we be without our plus (+) for addition, minus (−) for subtraction, × for multiplication, ÷ for division, and so on? We shall have more to say about these, and other, signs or symbols later.

Professor A. N. Whitehead, the eminent mathematical philosopher, writes in his *Introduction to Mathematics* (Home University Library)[1]:

> By relieving the brain of all unnecessary work, a good notation sets it free to concentrate on more advanced problems, and in effect increases the mental power of the race. Before the introduction of the Arabic notation, multiplication was difficult, and the division even of integers [whole numbers] called into play the highest mathematical faculties. Probably nothing in the modern world would have more astonished a Greek mathematician than to learn that, under the influence of compulsory education, a large proportion of the population of Western Europe could perform the operation of division for the largest numbers. . . . Mathematics is often considered a difficult and mysterious science, because of the numerous symbols which it employs. Of course, nothing is more incomprehensible than a symbolism which we do not understand. Also a symbolism, which we only partly understand and are unaccustomed to use, is difficult to follow. In exactly the same way the technical terms of any profession or trade are incomprehensible to those who have never been trained to use them. But this is not because they are difficult in themselves. On the contrary they have invariably been introduced to make things easy. So in mathematics, granted that we are giving any serious attention to mathematical ideas, the symbolism is invariably an immense simplification. It is not only of practical use, but is of great interest. For it represents an analysis of the ideas of the subject and an almost pictorial representation of their relations to each other. . . . By the aid of symbolism,

[1] Pp. 59, 60, 61.

we can make transitions in reasoning almost mechanically by the eye, which otherwise would call into play the highest faculties of the brain.

POWERS

A beautiful illustration of the use of a simple symbolism is afforded by what is called the index, or exponent, notation. In this notation a product like

$$2 \times 2 \times 2 \times 2 \ldots \text{(to 9 factors) is written as } 2^9,$$

the little 9, placed high and to the right of the 2, showing that nine factors 2 are to be multiplied together.

In the same way

$$10 \times 10 \times 10 \times 10 = 10^4.$$

We can express this in a more general way by saying that $a \times a \times a \times a = a^4$, where a represents any number. We owe this particular symbolism to Descartes, whom we have previously mentioned as the inventor of graphs. He made

RENÉ DESCARTES

one queer exception; he wrote $a \times a$ as aa, not as a^2, as we do.

In words we read 2^9 as 2-to-the-ninth, a^4 as a-to-the-fourth, and so on. But a^2 we call a-squared, and a^3 a-cubed. From our study of square numbers it is easy to see how a^2 came to be known as a-squared. Imagine a cube composed of small dots; the total number of the dots is obtained by multiplying together the number of dots in each of the three edges of the cube which meet at a corner—this will help you to remember a-cubed for a^3.

The little number placed high and to the right which tells us the number of factors to be multiplied together is called an *index* or an *exponent*. The number which is being multiplied is called the *base*. The resulting product is called a *power* of the base. Thus a^7 is the base a raised to the seventh power.

USING EXPONENTS

Consider these two sets of numbers:

1	2	3	4	5	6	7	8	9	10
2	4	8	16	32	64	128	256	512	1024

Nearly five hundred years ago it was noticed that the product of any two numbers in the lower row lay beneath the sum of the corresponding numbers in the upper row.

Thus $8 \times 32 = 256$, corresponding to
$3 + 5 = 8$ in the upper row.

This is an example of a remarkable property of exponents and powers of the same base. We can put it more generally thus:

$$a^3 = a \times a \times a \qquad \text{(3 factors)}$$
$$a^5 = a \times a \times a \times a \times a \text{ (5 factors).}$$

So $a^3 \times a^5 = a \times a \times a \ldots$ to $3 + 5$, or 8, factors. That is,

$$a^3 \times a^5 = a^{3+5} = a^8.$$

We can express this still more generally by writing $a^m \times a^n = a^{m+n}$, where m and n are any whole numbers. Can you argue in the same way that

$$a^5 \div a^3 = a^{5-3} = a^2?$$

And, again generally, that
$a^m \div a^n = a^{m-n}$, where m and n are whole numbers, and m is greater than n?

Suppose now that we divide a^5 by a^5. We make our

first big mathematical jump, or argument from the reasonableness of what we have just proved, by saying that

$a^5 \div a^5 = a^{5-5} = a^0$. (We call this applying the principle of continuity.) But we know that

$a^5 \div a^5 = 1$, and so we get
$a^0 = 1$.

This may be astonishing and disconcerting to anyone unaccustomed to any mathematics but ordinary arithmetic. For the present we leave the matter as just another example of the utility and convenience of a symbolic notation; later we shall see how this exponent-notation lies at the very root of the simplified calculations which the use of logarithms and the slide-rule afford us.

SQUARE ROOTS

We have seen that the square of a number is obtained by multiplying that number by itself; for example, the square of 10 is 10×10, or 100.

The reverse of this process—that is, finding that number which, multiplied by itself, is equal to a given number—is called finding the *square root* of the given number. For instance, the square root of 100 is 10, since 10 multiplied by 10 is equal to 100. We write this:

$$\sqrt{100} = 10.$$

The symbol $\sqrt{}$ is derived from a small r (for root) with a long tail, making $\frown$ or $\sqrt{}$.

The square roots of numbers, when they 'come out' exactly, can be found by trial. For example, to find $\sqrt{256}$ we proceed as follows:

Try $10 \times 10 = 100$. "Too small."
Try $20 \times 20 = 400$. "Too large."

32

Try $15 \times 15 = 225$. "Too small, but much closer."
Try $16 \times 16 = 256$. "Found!"

(The remainder of this section may be left, if desired, until decimals have been reached.)

Let us now examine the possibility of finding $\sqrt{8}$.

Try $2 \times 2 = 4$. "Too small."
Try $3 \times 3 = 9$. "Too large."

So $\sqrt{8}$ lies between 2 and 3.

We next multiply 8 by 100 or 10^2, and use the fact that $\sqrt{8} = \sqrt{\frac{800}{100}} = \sqrt{800} \div 10$. To find $\sqrt{800}$:

Try $20 \times 20 = 400$. "Too small."
Try $30 \times 30 = 900$. "Too large."
Try $25 \times 25 = 625$. "Still too small."
Try $28 \times 28 = 784$. "Still too small."
Try $29 \times 29 = 841$. "Too large."

So $\sqrt{800}$ lies between 28 and 29.

And so $\sqrt{8}$ lies between $\frac{28}{10}$ and $\frac{29}{10}$ (see above)—that is, between 2·8 and 2·9.

If further accuracy is desired we could next show that $\sqrt{8}$ lies between 2·82 and 2·83, and so on, to still further places of decimals. The inquiring reader will surely ask, "And how much further?" But that is another question, and the answer must have a place to itself later on.

Of course, there is a definite arithmetical method for finding square roots; and there are also more mechanical methods, depending ultimately on the exponents about which we have just been reading.

NUMBER-PATTERNS

We saw, at pp. 25–26, that the Greeks pictured certain numbers as having orderly designs, and that those numbers had definite properties which could be worked out from those designs.

There are other ways in which pattern or design in numbers can be seen and used. For example, multiples of 37 give us very striking results.

$$3 \times 37 = 111$$
$$6 \times 37 = 222$$
$$9 \times 37 = 333$$
$$\cdots\cdots\cdots\cdots$$

Do you see how the pattern is worked out, first on the right-hand side, then on the left-hand side? Write down the next four rows of the same pattern. Check two of them by actual multiplication.

It is clear that after we get to 999 on the right-hand side some change in the pattern must occur. Obtain 30×37, 33×37, 36×37 by actual multiplication, and see if you can recognize and continue the new pattern. If you are not already tired the next pattern change is at 57×37.

Here is another pattern based on 3367:

$$33 \times 3367 = 111,111$$
$$66 \times 3367 = 222,222$$
$$\cdots\cdots\cdots\cdots\cdots\cdots$$

Can you continue it?

Do you recognize the pattern in the following example?

$$1 \times 91 = 091$$
$$2 \times 91 = 182$$
$$3 \times 91 = 273$$
$$4 \times 91 = 364$$
$$\cdots\cdots\cdots\cdots$$

Can you continue it? Where does it break down?

Now, check for yourself that when 142,857 is multiplied by 2, 3, 4, 5, or 6, the result in each case has the same set of figures in a different order. Notice how they are arranged. Strange, isn't it?

Next,
$$11^2 = 121$$
$$111^2 = 12,321$$
$$1111^2 = 1,234,321$$
. . . .

The figures on the right-hand side are called *palindromic* figures, because they read the same whether taken forward or backward. Can you write down the next line? And the next line? At what square does the pattern fail?

Lastly, here are some tests by which you can check whether you have an *eye*—and possibly a *head*—for figures. In each case what are the missing figures in the spaces marked by the dots . . .? And what is the next line in the pattern for each set? No multiplication or division must be done, except possibly for checking.

$$1 \times 8 + 1 = 9 \qquad\qquad 9 \times 9 + 7 = 88$$
$$12 \times 8 + 2 = 98 \qquad\qquad 98 \times 9 + 6 = 888$$
$$123 \times 8 + 3 = 987 \qquad 987 \times 9 + 5 = 8888$$
$$1234 \times 8 + 4 = \ldots \qquad \ldots \times 9 + 4 = 88,888$$

$$28 \times 15,873 = 444,444$$
$$35 \times 15,873 = 555,555$$
$$42 \times 15,873 = 666,666$$
$$.. \times 15,873 = 777,777$$

$$1 \times 7 + 1 = 8 \qquad\qquad 11 - 2 = 9 \times 1$$
$$12 \times 7 + 2 = 86 \qquad\qquad 111 - 3 = 9 \times 12$$
$$123 \times 7 + 3 = 864 \qquad 1111 - 4 = 9 \times 123$$
$$1234 \times 7 + 4 = \ldots \qquad 11,111 - . = 9 \times \ldots$$

MOST NUMBERS TELL A STORY

Frequently in our work it is necessary to know if a number is exactly divisible by another number. One way of finding out is by actual division, but with some of the smaller divisors easy tests which do not involve complete division are possible. In what follows we shall use 'divisible by' to mean 'exactly divisible by.'

If the last figure of a number is even the number is divisible by 2. This is so obvious that it does not require proof.

If the last two figures of a number form a number which is divisible by 4 the number itself is divisible by 4.

The figures which are left when the last two figures are removed tell us the number of hundreds in the number; 100 is divisible by 4, so any number of 100's must be divisible by 4.

E.g., 38,276 = 38,200 + 76; 38,200 is 382 hundreds and so is divisible by 4. So we just have to inquire whether 76 is divisible by 4; this is so, and thus 38,276 can be divided by 4.

Test in the same way that 4,789,638 is not divisible by 4.

If the last three figures of a number form a number which is divisible by 8 the number itself is divisible by 8.

The figures which are left after the last three figures have been removed are thousands, and since $1000 \div 8 = 125$, any number of 1000's must be divisible by 8.

Thus 287,235 is not divisible by 8, since 235 is not divisible by 8.

But 19,672 is divisible by 8, since 672 is divisible by 8.

A number is divisible by 3 if the sum of its digits (that is, figures) is divisible by 3. A number is divisible by 9 if the sum of its digits is divisible by 9.

E.g., in 21,675 the sum of the digits is $2 + 1 + 6 + 7 + 5$—that is, 21; 21 is divisible by 3, but not by 9. So 21,675 is divisible by 3, but not by 9. On the other hand 34,785 is divisible by 9. Apply the rule yourself.

The proof in the general case is pretty, and for that reason we give it.

Suppose the number is

.....	thousands	hundreds	tens	units
	d	c	b	a

In words, reading it from right to left, it is a units, plus b tens, plus c hundreds, plus d thousands, and so on.

In symbols it is $a + 10b + 100c + 1000d + \ldots$

The sum of the digits is

$$a + b + c + d + \ldots$$

Subtracting the sum of the digits from the number, we have that the difference is

$$9b + 99c + 999d + \ldots$$

Now this difference is clearly divisible by 9 for any number of any size. If the sum of the digits is also divisible by 9 it follows that the number itself is also divisible by 9.

A number which ends in 5 or 0 is divisible by 5. A number which ends in 0 is divisible by 10, a number which ends in 00 is divisible by 100, and so on.

The verification of these facts is left to the reader.

MORE NUMBER-COMBINATIONS

Early in this chapter we discovered some interesting facts about *sums* of 'runs' of numbers starting from one. We shall now investigate some problems which depend upon *products* of 'runs' of numbers. Here is one:

It is possible to go from one place to another place by 8 different bus, tram, and train routes. In how many different ways can a person go by one route and return by another?

Let us suppose that the person chooses any one of the routes for his outward journey. Then for his return journey he has a choice of 7 different routes (why not

37

8?). But he can choose his outward route in 8 different ways, and to each of these ways there is a choice of return by 7 different routes. So altogether there are 8 × 7, or 56, ways of going by one route and returning by a different one. Scarcely credible, is it?

Now let us take a more personal example. A man has 3 hats, 2 overcoats, and 4 pairs of gloves. In how many ways can he select one each of these articles so that he can go out dressed differently each time?

He can select a hat in 3 ways. Then he can choose his overcoat in 2 ways. So he can select a hat and an overcoat in 3 × 2 ways. Similarly, he has 4 choices for his pair of gloves, so altogether he can choose the 3 articles in 3 × 2 × 4, or 24, ways.

One more numerical example:

In how many different ways can the 6 colours in snooker be placed one in each of the 6 pockets of a billiards table?

For the first colour we have a choice of 6 pockets; then the second has to go into one of the other 5 pockets. So the first two colours can be placed in pockets in 6 × 5 different ways. The third ball has a choice of 4 pockets; so the first three balls can be distributed in the pockets in 6 × 5 × 4 ways. Similarly, four balls can be disposed of in 6 × 5 × 4 × 3 ways; five balls in 6 × 5 × 4 × 3 × 2 ways; and the sixth ball goes into the remaining pocket—in one way. So we can write down the number of ways of placing the six different colours in the six pockets as 6 × 5 × 4 × 3 × 2 × 1, which on multiplication gives 720.

There is a simple way of writing down these products of consecutive numbers starting from 1. We write 6 × 5 × 4 × 3 × 2 × 1 as 6! (or ⌊6, as it is sometimes written), the exclamation mark after the 6 denoting that all the whole numbers 1 to 6 are to be multiplied together. We call it *factorial* six. In the same way

$$4! = 4 \times 3 \times 2 \times 1$$
and $n! = n \times (n - 1) \times (n - 2) \times \ldots \times 3 \times 2 \times 1.$

Of course we cannot evaluate $n!$ any farther unless we know what the numerical value of n is.

It is easy to see that the number of ways in which the 26 letters of the alphabet can be made up into different words of 26 letters is 26! (Of course the words so formed would make no sense.) Think what a saving of labour of writing this factorial notation, or symbolism, affords.

EXERCISE 1

1. Write down the odd numbers, in order, up to 50. By using Eratosthenes's sieve find all the prime numbers less than 50.

2. The squares in Fig. 4 can each be fitted together to form a large square. They illustrate:

$$1 + 3 = 2^2.$$
$$1 + 3 + 5 = 3^2.$$
$$1 + 3 + 5 + 7 = 4^2.$$

Can you continue the process to show 5^2, 6^2, etc.?
What do the squares tell you about the sums of all the odd numbers in order, each starting from 1?

FIG. 4

3. Express in simpler form: $a^4 \times a^3$; $3^6 \times 3^2$; $10^3 \times 10^5$; $2^2 \times 2^2 \times 2^2$; $5^3 \times 5^4 \times 5$; $x^8 \div x^5$; $7^6 \div 7^5$; $a^4 \times a^7 \div a^5$; $2^2 \times 2^3 \div 2^5$; $10^8 \times 10^7 \div 10^9$.

4. Find the (exact) square roots of these numbers by trial: 169; 361; 576; 324; 729.

5. Find two whole numbers between which the square roots of these numbers lie: 200; 150; 234; 444.

6. Find which of these numbers are divisible by 4, 5, 9, 10, 6, 8, 12 (no actual division of the numbers must be performed): 435; 690; 1437; 2648; 3332; 2565; 123,456; 1,223,748.

7. In how many ways can five people sit on a bench?

8. How many numbers of four digits can be formed from the five digits 1, 2, 3, 4, 5 if none of the digits may be repeated?

9. In how many ways can six people pair off for a duet?

10. (Harder.) In how many ways can five people be arranged round a circular dinner table?

11. (Harder.) How many numbers of four digits and greater than 4000 can be formed from the digits 2, 3, 4, 6, 8, 9? No digit is to be repeated in a number.

12. (Still harder.) In how many ways can six men and six ladies sit at a round table so that the men and the ladies sit in alternate seats?

CHAPTER III

A NEW KIND OF NUMBER

PRELIMINARY CONSIDERATIONS

SUPPOSE you were asked to evaluate $7 + 5$. You would say to yourself: I have to add 5 to 7. Almost without realizing it, you regard the symbol 'plus' as a direction to do something; it is an instruction to add. In the same way $7 - 5$ means that 5 is to be taken away from 7. The 'minus' may be regarded as an instruction to subtract.

Now suppose that you have 8 pennies in your pocket. If you give away 5 pennies there are still 3 pennies left in your pocket; if 8 pennies are given away your pocket is empty. But you cannot give away 10 pennies; nor can you take away 10 pennies from 8 pennies.

Suppose now you take 8 paces forward. If you then step back 3 equal paces, still facing in the forward direction, you will still be 5 paces in front of the place from which you started. If you had stepped back 8 equal paces you would have found yourself in the place from which you started. If you had stepped 10 paces backward you would have found yourself 2 paces behind the place from which you started. The same form of argument could be used if you were standing near the middle of a flight of stairs, and took steps up and down, without turning round, on the stairs.

How can we write down, in figures, numbers which will tell us at once such things as whether we are 3 paces in front or behind, 3 steps above or below, our starting-point? It could perhaps be done by our writing 3 paces to mean 3 paces forward, and Ɛ (3 written backward) to indicate 3 paces backward. But this would be very awkward—and what could be done with the figure 8?

Actually we make use of the plus and minus signs. If we agree to write +3 paces to show 3 paces forward then we can write −3 paces to show the opposite—that is, 3 paces stepped backward. But note, the + and − signs here are not instructions to add or to subtract. They tell us something about the kind of paces we take —not their length, but whether they are taken forward or backward. Used in this way, the plus and minus signs may almost be regarded as adjectives.

Here are some examples of the use of + and − in this way:

If + 5° means a temperature of 5° above zero, − 5° means a temperature of 5° below zero.

If + 20 miles shows 20 miles north of a place, − 20 miles shows 20 miles south of it.

If + 200 ft. shows 200 ft. measured vertically upward, − 200 ft. shows 200 ft. vertically downward.

If + £10 shows £10 gain (or money in the bank), − £10 shows £10 loss (or overdraft).

If + 5 weeks means 5 weeks hence, − 5 weeks means 5 weeks ago.

These numbers which are not complete without the appropriate signs in front of them are called *signed numbers*, or *directed numbers*; those with a + sign in front of them are called *positive numbers*; those with a − sign in front are called *negative numbers*.

It is a point of great historic interest that in olden days scholars did not believe that such things as negative numbers could exist, and this held up mathematical progress for a very long time.

POSITIVE AND NEGATIVE NUMBERS

Study the scale in Fig. 5.

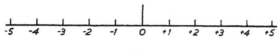

FIG. 5

You will see that on the right of the zero (0) the positive numbers + 1, + 2, + 3, etc., are marked. Now look to the left of the zero. There the negative numbers are marked − 1, − 2, − 3, etc., from zero, but in this case towards the left. It is easy to see how this fits in with our introduction to signed numbers by means of paces taken. For if we take going to the right as the positive direction we may think of a movement from 0 to + 3 on the scale as 3 paces to the right; 3 paces to the left will then be a shift from 0 to − 3 on the scale.

We have now a scale which extends in both directions, and as far as we please from zero, + to the right, − to the left. In the same way we can think of a scale which is marked with positive numbers upward from 0 and with negative numbers downward from 0 (Fig. 6). The positive, or plus, numbers are the same as the ordinary numbers of arithmetic; they are all greater than zero. If a number is written without a sign in front of it we consider it a positive number if the necessity ever arises. Negative numbers are less than zero; the sign of a negative number must *always* be written before it.

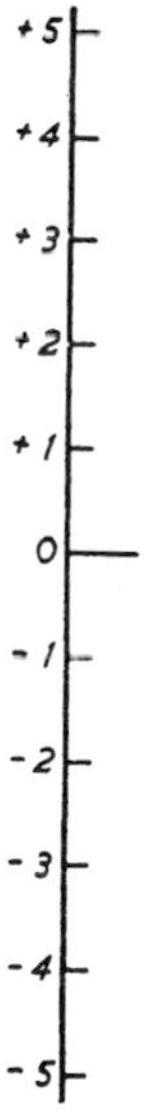

FIG. 6

If the reader cares to think further on this subject of positive and negative numbers he will realize that

43

whether a number is positive or negative depends upon where the zero is fixed. To take a simple illustration: if we consider the stairs of a house the first-floor landing may be taken as the zero mark. Upward to the attic is then positive, and downward to the ground floor is negative. But if the zero mark is taken at ground-floor level, up to the first floor then becomes positive. A mischievous person might be tempted to ask: what becomes of your two-way scale if you take the cellar as zero level?

All this discussion of signed numbers might appear at first sight to be rather academic, but signed numbers are of great use in many ways. It has already been hinted that mathematics would have been at a standstill were it not for the introduction of negative numbers. The interpretation of a negative sign in the solution to a problem can often tell a trained mathematician what he wants to know. Practically, when figures are to be compared the use of the $+$ and $-$ tells at a glance what has happened. Look at the Stock Exchange prices in the daily papers, or at farmers' records, population figures, etc. The $+$ and $-$ tell at a glance what it is desired to convey about increases or decreases. (No doubt many people would prefer their bankers to used signed numbers in indicating the state of their balances. They would not provide such a shock as those figures written or typed in red.)

ADDING AND SUBTRACTING WITH SIGNED
NUMBERS

We have to be careful in operations involving the addition and subtraction of signed numbers, as we have to use the $+$ and $-$ signs both to indicate signed numbers and as instructions to add or subtract. We simplify matters by enclosing the numbers, with their appropriate signs, within parentheses or brackets ( ) when there is any possibility of confusion.

44

The method of *addition* may be appreciated by a study of the following examples:

(i) £5 gain plus £7 gain makes £12 gain. In symbols:
$$+ 5 + (+ 7) = + 12.$$

(ii) £5 gain plus £5 loss makes nothing gained or lost. In symbols:
$$+ 5 + (- 5) = 0.$$

(iii) £5 loss plus £7 loss makes £12 loss. In symbols:
$$- 5 + (- 7) = - 12.$$

(iv) £5 loss plus £7 gain makes £2 gain. In symbols:
$$- 5 + (+ 7) = + 2.$$

(v) £5 gain plus £7 loss makes £2 loss. In symbols:
$$+ 5 + (- 7) = - 2.$$

Brackets have been placed around the second number only in each case, for only there can confusion arise whether the sign is operative as a verb or an adjective.

The method may also be deduced from carrying out similar operations on the first number-scale given at p. 43. The result of the operations should be illustrated in symbols as above.

(*a*) Start at + 2, then go 3 to the right.
(*b*) Start at + 2, then go 3 to the left.
(*c*) Start at − 3, then go 5 to the right.
(*d*) Start at   0, then go 4 to the left.
(*e*) Start at − 2, then go 3 to the right, then 2 to the left.
(*f*) Start at + 4, then go 8 to the left, then 4 to the right.

From these exercises it should appear that we can put the matter generally as follows:

When adding numbers with the same signs find their sum

(*as in arithmetic*), *and put in front of it the same sign as that which all the numbers have.*

When adding numbers with different signs find the sum of the positive numbers, and then find the sum of the negative numbers. Subtract the smaller sum from the larger sum, and put in front of the answer the sign of the larger sum. If the sign is a plus you may leave it out.

For example,

$$(+ 7) + (- 12) + (+ 14) + (- 6)$$
$$= (+ 21) + (- 18) = (+ 3), \text{ or just } 3.$$
$$(+ 7) + (- 12) + (- 14) + (+ 6)$$
$$= (+ 13) + (- 26) = (- 13), \text{ or just } - 13.$$

The method of *subtraction* may be deduced from the following examples. Mr Bright and Mr Jones are two business men:

(i) If Mr B. makes £7 and Mr J. makes £5, Mr B. is £2 better off than Mr J. In symbols:
$$+ 7 - (+ 5) = + 2.$$

(ii) If Mr B. makes £5 and Mr J. makes £7, Mr B. is £2 worse off than Mr J. In symbols:
$$+ 5 - (+ 7) = - 2.$$

(iii) If Mr B. makes £7 and Mr J. loses £5, Mr B. is £12 better off than Mr J. In symbols:
$$+ 7 - (- 5) = + 12.$$

(iv) If Mr B. loses £7 and Mr J. loses £5, Mr B. is £2 worse off than Mr J. In symbols:
$$- 7 - (- 5) = - 2.$$

From this you should be able to verify the general statement:

When a signed number is subtracted from another signed number it is sufficient to change its sign and add. Thus:

$$+ 7 - (+ 5) = + 7 + (- 5) = + 2.$$
$$+ 5 - (+ 7) = + 5 + (- 7) = - 2.$$
$$+ 7 - (- 5) = + 7 + (+ 5) = + 12.$$
$$- 7 - (- 5) = - 7 + (+ 5) = - 2.$$

MULTIPLYING AND DIVIDING WITH SIGNED NUMBERS

We shall next show how multiplication and division are done when signed numbers are involved. Again we take practical examples from which we can deduce the general results:

(i) If a man saves 8s. per week, then in five weeks he will have 40s. more than he has now. In symbols:

$$(+ 5) \times (+ 8) = + 40.$$

(ii) If he spends 8s. per week more than he earns, then in five weeks he will have 40s. less than he has now. In symbols:

$$(+ 5) \times (- 8) = - 40.$$

(iii) If he saves 8s. per week, then five weeks ago he had 40s. less than he has now. In symbols:

$$(- 5) \times (+ 8) = - 40.$$

(iv) If he spends 8s. per week more than he earns, then five weeks ago he had 40s. more than he has now. In symbols:

$$(- 5) \times (- 8) = + 40.$$

Stating our results more generally, we may say:

The result of multiplying a positive number by a negative number, or a negative number by a positive number, is negative.

The result of multiplying together two positive numbers, or two negative numbers, is positive.

In each case the number-value of the product is the same as is obtained when the numbers are multiplied together, without regard to the signs attached to them.

47

Here are some sketches which illustrate the same points. We shall suppose that Mr Hustler's paces are each 3 ft. long; that the positive (or plus) direction is to the right; that 'step back' means negative (or minus) lengths to the paces; and that the command "about turn" changes the direction of the paces—that is, changes $+2$ paces into -2 paces, and so on.

Two paces forward, march (Fig 7).

Two paces step back, march (Fig 8).

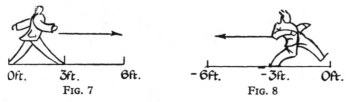

Fig. 7   Fig. 8

About turn. Two paces forward, march (Fig 9).

About turn. Two paces step back, march (Fig 10).

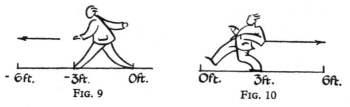

Fig. 9   Fig. 10

By reversing the arguments we can show that similar deductions may be made in respect of division where signed quantities are involved. That is:

The result of dividing a positive number by a negative number, or a negative number by a positive number, is negative; and the result of dividing a positive number by a positive number, or a negative number by a negative number, is positive. For example:

$$\frac{+12}{-3} = -4; \quad \frac{-20}{+4} = -5; \quad \frac{+36}{+9} = +4; \quad \frac{-18}{-6} = +3.$$

EXERCISE 2

1. Which of these would you show with + signs and which with − signs?

   (a) Received £10; spent £5.
   (b) Received £4; received £6.
   (c) Spent 5s.; spent 4s.
   (d) Temperature rises 14°; falls 8°.
   (e) Lift goes down 250 ft.; rises 100 ft.
   (f) River level falls 8 ft.; rises 7 ft.

In each case represent the result of the two operations symbolically, and interpret your results.

2. The Allies landed in Normandy on D-day, June 6, 1945. When were D-day + 3; D-day + 17; D-day + 30; D-day −1; D-day −15?

3. Interpret these statements:

   (a) The height above sea-level of the bottom of a well is −25 ft.
   (b) In a race Charles had a start of −8 ft.
   (c) A temperature of −4° Centigrade.
   (d) A speed of −20 m.p.h.

4. Give the balances: $(+12) + (-6)$; $(-5) + (-17)$; $(-16) + (+14) + (-20)$; $(+12) - (+7)$; $(+42) - (-8)$; $(-8) - (+26)$; $(-16) + (-16) - (-32)$; $(-25) - (-52) - (+37) + (+19)$.

5. Work out: $(+4) \times (-13)$; $(-3) \times (-4)$; $(-8) \times (+9)$; $(-3) \times (-4) \times (-5)$; $(20) \div (-5)$; $(-32) \div (-8)$; $(-72) \div (+12)$.

6. Write down the squares of: -4; -7; -9; -12; -15; and the cubes of -3; -4; -6.

CHAPTER IV

FRACTIONS

Multiplication is vexation,
Division's twice as bad,
The rule of three perplexes me,
And fractions drive me mad.

DID *you* have trouble with fractions in arithmetic? It is a common experience, and many have felt that the composer of the old jingle quoted above echoed their thoughts.

The history of mathematics shows that the human race found fractions difficult and gained an understanding of them very slowly. For many centuries only very simple fractions were employed. The ancient Egyptians used the fraction $\frac{1}{2}$, but instead of writing $\frac{3}{4}$ they wrote $\frac{1}{2} + \frac{1}{4}$. The Romans reduced all their fractions to twelfths, and for $4\frac{1}{4}$ feet wrote 4 feet 3 twelfths. (We owe it to the Romans that we have twelve pence to a shilling, twelve inches to a foot, and twelve ounces to the pound Troy still used by chemists.) Of course the advantage of 12 is that it has factors 2, 3, 4, and 6; 10, on the other hand, only has the factors 2 and 5. The Roman sub-divisions—that is, their twelfths—were all called *unciæ*, a word from which our 'inch' and 'ounce' are derived.

It is a fact that through all ages men have divided larger units of weights and measures into smaller ones— *e.g.*, pounds into ounces and yards into feet—in order not to be bothered with fractions of the larger units.

It seems strange now to reflect what a big step it was

to proceed from the idea of a fraction like $\frac{1}{4}$ (with 1 in the top line) to a fraction like $\frac{3}{4}$ (with a number other than 1 in the top line); and that it probably took many centuries to make it. Indeed, it took a still longer time to develop a satisfactory symbol for a fraction itself, as we shall see presently.

After this you will not be surprised to read that rules for handling fractions did not appear until about A.D. 1000. So if you find that "fractions drive you mad" you may take comfort from the fact that it took nearly three thousand years to develop the rules for dealing with them. The idea of extending fractions to include such expressions as $\frac{2}{2}$ and $\frac{3}{2}$ did not occur to arithmeticians till quite modern times.

WHAT IS A FRACTION?

From the very beginning a fraction has simply been regarded as a part of something, and so—something broken off. Thus our word fraction is derived from the Latin verb *frangere*, to break. The first arithmetic book in English, written by Robert Recorde in 1540, took the form of a dialogue between master and scholar. This is what Recorde's scholar thought a fraction was:

Marry sir, I think a Fraction (as I have heard it often named) to be a

ROBERT RECORDE

broken number, that is to say to be no number but a part of a number.

51

Incidentally, the name 'vulgar fractions' goes back to about the same time. The fractions in everyday use were sometimes called *minutiæ vulgares*, or 'the usual small fractions'; hence our 'vulgar' fractions—not, as was stated in the biography of a famous lady published some years ago, because they were not fit to be taught in very genteel girls' schools.

But to return to our question—what is a fraction? Take, for example, ¾—in words, three-quarters or three-fourths. We can attach two meanings to this. Perhaps they will better be explained diagramatically.

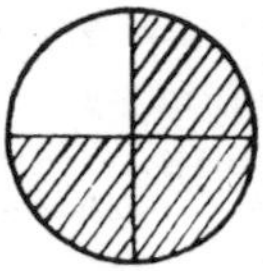

(*a*) Draw a circle. Divide it into four equal parts by making two cuts through the centre of the circle, as in Fig. 11. Then the shaded part represents three-quarters of the circle—that is, three of the four equal parts into which the circle is divided.

FIG. 11

(*b*) Now draw three equal circles, and divide each into four equal parts as before. Then each shaded part represents one-quarter of each circle (Fig. 12). So three of the shaded parts, which we have seen represent three-quarters of a circle, together make one-quarter of three circles. That is, ¾ also means 3 divided by 4.

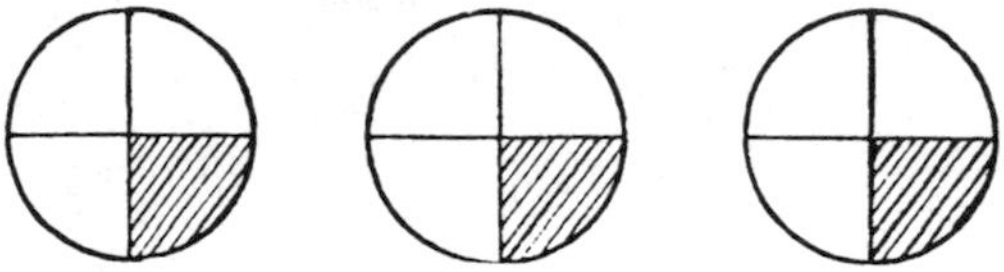

FIG. 12

The top line (here 3) of the fraction is called the numerator; the bottom line (here 4) is called the denominator. So we see that

$$a\ fraction = \frac{numerator}{denominator}, \text{ and also}$$

a fraction = numerator divided by the denominator.

Historical note. The development of an adequate notation for fractions exercised the minds of scholars for centuries. Diophantus (a Greek of about A.D. 275) was one of the first to approximate to our present notation. He wrote what we should call the denominator *above* the numerator. This would be equivalent to having $\frac{4}{3}$ (without the dividing bar) stand for our $\frac{3}{4}$. The system of writing the numerator *above* the denominator was first used by the Hindu Brahmagupta in the seventh century A.D. The Arabs improved on this by putting in a bar between the two numbers.

To return to our argument: from Fig. 11 we also see that if the numerator is less than the denominator the fraction is less than 1. Such fractions are called *proper* fractions.

If the numerator is equal to the denominator the fraction is equal to 1.

If the numerator is greater than the denominator we must look farther into the matter. Take, for example, $\frac{5}{4}$. This means five-quarters, or the shaded portions of

FIG. 13

Fig. 13. That is, $\frac{5}{4}$ is greater than 1. So if the numerator of a fraction is greater than the denominator the fraction is greater than 1. We call such fractions *improper* fractions.

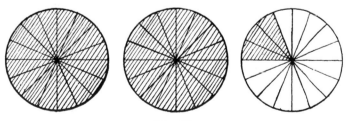

FIG. 14

One more explanation, and we shall be ready to proceed. Examine the three equal circles of Fig. 14, which are each divided into sixteen parts, or sixteenths.

53

The shaded parts show two whole circles and three-sixteenths of one circle—that is, $2\frac{3}{16}$ circles. But we may also say that thirty-five sixteenths of a circle have been shaded. So:

$$2\tfrac{3}{16} = \tfrac{35}{16}.$$

A number like $2\frac{3}{16}$ is called a *mixed number*. It is a mixture of a number and a fraction. It is a short form of $2 + \frac{3}{16}$ with the $+$ omitted.

Note. The whole number in a mixed number was originally placed to the right of the fractional part, and not to the left, as we put it.

Of course, in practice we do not have to draw circles in order to change a mixed number into an improper fraction. We merely say: multiply 2 by 16, add 3, and place the 35 so obtained over 16. Similarly

$$7\tfrac{3}{8} = \tfrac{59}{8} \ (7 \times 8 = 56, \text{ add } 3, \text{ making } 59).$$

We can change an improper fraction to a mixed number by reversing the process. Let us consider $\frac{17}{6}$—that is, seventeen-sixths. If 17 is divided by 6 we have 2 plus a remainder of 5 (in terms of circles, divided into sixths, two whole circles and five shaded parts out of six in a third circle). But this remainder of five does not mean the whole number five, but five-sixths. So:

$$\tfrac{17}{6} = 2\tfrac{5}{6}.$$
$$\text{Similarly } \tfrac{53}{10} = 5 + \text{three-tenths} = 5\tfrac{3}{10}.$$

Note. We use much the same idea when a measure or a sum of money has to be distributed in equal shares among several people. Thus if five guineas (*i.e.*, 105*s.*) has to be distributed equally among eleven people each person has 9*s.* and there is a remainder of 6*s.* This 6*s.*, converted into 72*d.*, allows each person to have another 6*d.*, and there is 6*d.* left over. Here the *d.* after the 6 indicates the nature of the remainder.

PROCESSES WITH FRACTIONS

We now touch upon the application to fractions of the four fundamental operations—addition, subtraction, multiplication, and division. Elementary as they may seem to some of us, these operations were for many hundreds of years a stumbling-block to further progress. Addition was indeed vexation, but when it came to multiplication early mathematicians were sorely perplexed, because after a number was *multiplied* by a fraction, did it not become smaller than before, and not larger, as was their experience when multiplying by whole numbers?

Eventually the problem of addition and subtraction was solved by the idea of equivalent fractions; so to equivalent fractions we must next turn.

EQUIVALENT FRACTIONS

Examine the square drawn in Fig. 15.

It is divided into four large squares, or a hundred small squares. Now the two upper large squares can be regarded as either:

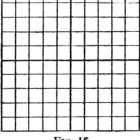

(*a*) one-half ($\frac{1}{2}$) of the whole square; or

(*b*) two-fourths, or two-quarters ($\frac{2}{4}$)—that is, two large squares out of the four large squares that make up the whole square; or

Fig. 15

(*c*) fifty-hundredths ($\frac{50}{100}$)—that is, fifty small squares out of the hundred small squares which make up the whole square.

But $\dfrac{2}{4} = \dfrac{1 \times 2}{2 \times 2}$, and $\dfrac{50}{100} = \dfrac{1 \times 50}{2 \times 50}$; so we see that the numerator and the denominator of a fraction can each be multiplied by 2, or by 50—or, in fact, by any other number—without altering the value of the fraction.

Similarly, we can take fifty-hundredths ($\frac{50}{100}$) and divide the numerator and the denominator each by 10 to get $\frac{5}{10}$, or each by 25, to get $\frac{2}{4}$, or each by 50, to get $\frac{1}{2}$.

This may be written:

$$\frac{50}{100} = \frac{\cancel{10} \times 5}{\cancel{10} \times 10} = \frac{5}{10}; \qquad \frac{50}{100} = \frac{\cancel{25} \times 2}{\cancel{25} \times 4} = \frac{2}{4};$$

$$\frac{50}{100} = \frac{\cancel{50} \times 1}{\cancel{50} \times 2} = \frac{1}{2}.$$

This dividing out by 10, or 25, or 50, as the case may be, is called *cancelling*, the sloping lines indicating that the numbers which they cover have been cancelled out. When no further cancellation can take place a fraction is said to be its *lowest terms*. Thus $\frac{1}{2}$ is in its lowest terms; $\frac{2}{4}$, $\frac{5}{10}$, and $\frac{50}{100}$ are not. Those fractions which by cancellation can be shown to be equal are called *equivalent fractions*.

As a matter of fact, work in fractions often depends upon changing a fraction which is in its lowest terms to an equivalent fraction which is not in its lowest terms. This we shall see very shortly.

EXERCISE 3

1. By cancelling, obtain the following fractions in their lowest terms:

$$\frac{4}{6}, \quad \frac{9}{12}, \quad \frac{8}{12}, \quad \frac{15}{20}, \quad \frac{42}{56}, \quad \frac{16}{80}, \quad \frac{25}{100}, \quad \frac{55}{88}, \quad \frac{25}{40}, \quad \frac{36}{54}.$$

2. Pick out the equivalent fractions in the following groups:

$$\frac{4}{10}, \quad \frac{8}{14}, \quad \frac{20}{50}; \qquad \frac{9}{12}, \quad \frac{45}{60}, \quad \frac{20}{30}; \qquad \frac{18}{24}, \quad \frac{24}{36}, \quad \frac{36}{54}.$$

3. Fill in the spaces in the following:

$$\frac{1}{2} = \overline{24}, \qquad \frac{2}{3} = \overline{30}, \qquad \frac{5}{8} = \frac{25}{\quad},$$

$$\frac{7}{10} = \overline{100}, \qquad \frac{3}{16} = \overline{80}.$$

L.C.M.

Mr A, Mr B, Mr C, and Mr D are four commercial travellers. They happen to be staying at the same hotel one day. Mr A stays there every 4 days, Mr B every 6 days, Mr C every 8 days, and Mr D every 12 days. In how many days will they all be there together again?

We can solve this problem in the following way:

Mr A will be there in 4, 8, 12, 16, 20, 24, 28, .. days.
Mr B will be there in 6, 12, 18, 24, 30, ... days.
Mr C will be there in 8, 16, 24, 32, ... days.
Mr D will be there in 12, 24, 36, ... days.

Looking over these figures, we see that the first occasion on which they will all be together again is in 24 days. (Why the *first* occasion?)

We call this process *finding the lowest common multiple*, or *finding the L.C.M.*, of the numbers 4, 6, 8, 12. We have found the lowest number into which each of the numbers 4, 6, 8, 12 divides exactly.

All L.C.M.'s can be found in the manner shown above, but in practice the old-fashioned way is probably still the best. Here it is:

1. First write down the numbers, spacing them well apart.

2. Strike out any number which is contained in another of the numbers (here 4). Why?

3. 2 divides 6, 8, and 12 exactly. Perform the division in the usual way.

$$
\begin{array}{r|l}
2 & 4,\ 6,\ 8,\ 12 \\
\hline
2 & 3,\ 4,\ 6 \\
\hline
 & 2,\ 3
\end{array}
$$

57

4. Strike out the resulting 3. Why?

5. Again divide by 2. No further division is possible.

6. The L.C.M. is read as the product of all the numbers that remain at the side and in the last line—here $2 \times 2 \times 2 \times 3$, or 24.

Here is a more involved example. Note that a number which is not exactly divisible by the divisor (the number by which we are dividing) is just written down again.

$$
\begin{array}{r|rrrr}
2 & 2, & 15, & 18, & 24 \\
\hline
3 & & 15, & 9, & 12 \\
\hline
& & 5, & 3, & 4 \\
\hline
\end{array}
$$

$$\text{L.C.M.} = 2 \times 3 \times 5 \times 3 \times 4 = 360.$$

HOW TO ADD AND SUBTRACT FRACTIONS

We are at last in a position to add and subtract fractions. Let us start with $\frac{1}{3} + \frac{1}{4}$. We cannot add them directly any more than we can say that 3s. plus 4d. is 7s. or 7d. But if they are each changed to equivalent fractions having the same denominator the way is clear. This is where L.C.M.'s help us. For the L.C.M. of 3 and 4 is 12, and we can express each fraction as twelfths. Thus:

$$\frac{1}{3} + \frac{1}{4} = \frac{4}{12} + \frac{3}{12} = \frac{7}{12},$$

since three-twelfths added to four-twelfths is seventwelfths. In practice we generally write this as

$$\frac{1}{3} + \frac{1}{4} = \frac{4+3}{12} = \frac{7}{12},$$

the 4 plus 3 being written over the single denominator 12 to save time and trouble.

In the same way:

$$\frac{2}{3} - \frac{1}{6} = \frac{4-1}{6} = \frac{3}{6} = \frac{1}{2},$$

on cancelling down $\frac{3}{6}$ to its lowest terms.

Of course it is not always quite as easy as this, so a few examples are next given of some of the tricky points that may arise.

(a) $2\frac{3}{4} + 6\frac{5}{8}$

$= 8 + \frac{3}{4} + \frac{5}{8}$ (we first add the whole numbers)

$= 8 + \frac{6+5}{8}$

$= 8 + \frac{11}{8} = 8 + 1\frac{3}{8}$ (changing $\frac{11}{8}$ to a mixed number)

$= 9\frac{3}{8}.$

(b) $18\frac{1}{6} - 7\frac{3}{4}$

$= 11 + \frac{1}{6} - \frac{3}{4}$ (we first subtract the whole numbers)

$= 10 + 1\frac{1}{6} - \frac{3}{4}$ (we cannot subtract $\frac{3}{4}$ from $\frac{1}{6}$, so we write $11 + \frac{1}{6}$ as $10 + 1\frac{1}{6}$)

$= 10 + \frac{7}{6} - \frac{3}{4}$

$= 10 + \frac{14-9}{12} = 10 + \frac{5}{12} = 10\frac{5}{12}.$

(c) $4\frac{2}{3} - 3\frac{7}{8} + 6\frac{1}{4}$

$= 7 + \frac{2}{3} - \frac{7}{8} + \frac{1}{4}$

$= 7 + \frac{16-21+6}{24} = 7 + \frac{22-21}{24} = 7\frac{1}{24}.$

EXERCISE 4

1. $\frac{1}{2} + \frac{1}{3}$; $\frac{2}{3} + \frac{1}{4}$; $\frac{5}{6} + \frac{2}{3}$; $\frac{1}{2} + \frac{1}{3} + \frac{1}{4}$; $\frac{2}{3} + \frac{3}{4} + \frac{7}{12}.$

2. $\frac{1}{2} - \frac{1}{3}$; $\frac{2}{3} - \frac{1}{4}$; $2\frac{5}{6} - 1\frac{2}{3}$; $2\frac{1}{2} - 1\frac{3}{4}$; $8\frac{1}{10} - 2\frac{4}{15}.$

3. $4\frac{1}{6} + 7\frac{2}{3} + 5$; $7\frac{1}{4} + 3\frac{5}{6} + 4\frac{2}{3}$; $9\frac{1}{2} - 4\frac{3}{8} + 2\frac{1}{4}$; $12\frac{1}{6} + 3\frac{1}{2} - 9\frac{1}{3}.$

THE IDEA OF MULTIPLICATION AND DIVISION WHEN FRACTIONS ARE INVOLVED

Before we proceed with this section the point must be stressed that mathematics is an exact science, and because of this a mathematician has always to be careful that he knows exactly what he is given, what he is doing, and, if possible, what sort of result he wishes to get. So in mathematics we define what we mean by a word or phrase in a more exact sense than is customary, or perhaps even necessary, in everyday speech. Thus "Old So-and-so is infinitely richer than Mr Such-and-such" would pass unnoticed in general conversation; but 'infinity' and 'infinite' are words which, as we shall see, have to be used most carefully by the mathematician.

So it is with the simple words 'multiply' and 'divide.' We may read their cattle were multiplied in Gilead, or "they were divided up into three parties"; in the first case we associate *increase*, in the second *decrease*.

But when we come to fractions, what can we say? What is the meaning of $15 \times \frac{1}{3}$, or $2 \div \frac{1}{8}$?

Here the necessary precision of the mathematician comes to our aid. We agree to regard (in other words, we define) $15 \times \frac{1}{3}$ as $\frac{1}{3}$ of 15. With whole numbers 'of' has the significance of multiplication, and so, by regarding the $\times$ sign where fractions are involved as meaning 'of,' we are able to attach a reasonable meaning to the process of multiplication at the same time as we come to regard the operation itself as perfectly mechanical.

There may be less trouble in picturing the meaning of multiplication by a mixed number, but in effect the same implications are present as when the multiplication is by a fraction.

The process of division where fractions are involved is regarded, as in the case of division by whole numbers, as the reverse of the multiplication process.

Thus $2 \div \frac{1}{8} = 2 \times \frac{8}{1} = 16$, as is otherwise evident, for $\frac{1}{8}$ is contained 8 times in 1, and so 16 times in 2.

THE JUSTIFICATION FOR THE MULTIPLICATION PROCESS

Suppose we have to multiply $1\frac{3}{4}$ by $1\frac{1}{3}$.

The method is to change each mixed number into an improper fraction; the product is a fraction whose numerator is the product of their numerators and whose denominator is the product of their denominators.

Thus:

$$1\frac{3}{4} \times 1\frac{1}{3} = \frac{7}{4} \times \frac{4}{3} = \frac{28}{12} = 2\frac{4}{12} = 2\frac{1}{3}.$$

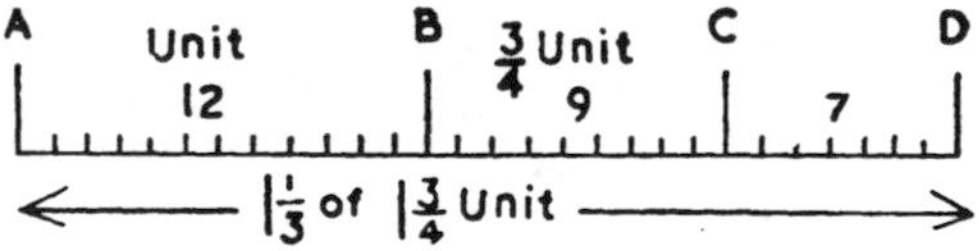

FIG. 16

Draw a straight line AB and divide it into 12 equal parts (4×3). A suitable length will be twelve-eighths of an inch.

We shall find $1\frac{1}{3}$ of $1\frac{3}{4}$ (of AB), and we shall regard AB as our unit.

If AB is extended to C so that $BC = \frac{3}{4} AB = 9$ divisions, then $AC = 1\frac{3}{4} AB = 1\frac{3}{4}$ of a unit; and $AC = 12 + 9 = 21$ divisions.

Now extend AC to D, so that $CD = \frac{1}{3} AC = \frac{1}{3}$ of $21 = 7$ divisions.

Then $AD = 1\frac{1}{3} AC = 1\frac{1}{3} \times 1\frac{3}{4}$ of a unit.

Also, $AD = 21 + 7 = 28$ divisions.

So $AD = \frac{28}{12}$ of a unit $= \frac{7}{3}$ of a unit, on cancelling down,
$$= 2\frac{1}{3} \text{ of a unit.}$$

That is, $1\frac{1}{3} \times 1\frac{3}{4}$ of a unit $= 2\frac{1}{3}$ of a unit,

or $1\frac{1}{3} \times 1\frac{3}{4} = 2\frac{1}{3}$.

We have seen (at p. 56) that a fraction may be brought to its lowest terms by cancelling. The process of multiplication may be simplified greatly by *cross-cancelling*.

Study the following examples:

(a) $\quad \dfrac{10}{27} \times \dfrac{3}{4} = \dfrac{\overset{5}{\cancel{10}}}{\underset{9}{\cancel{27}}} \times \dfrac{\overset{1}{\cancel{3}}}{\underset{2}{\cancel{4}}} = \dfrac{5}{9} \times \dfrac{1}{2} = \dfrac{5}{18}.$

(b) $\quad 2\dfrac{4}{7} \times 5\dfrac{1}{3} \times 2\dfrac{5}{8} = \dfrac{\overset{6}{\cancel{18}}}{\underset{1}{\cancel{7}}} \times \dfrac{\overset{2}{\cancel{16}}}{\underset{1}{\cancel{3}}} \times \dfrac{\overset{3}{\cancel{21}}}{\underset{1}{\cancel{8}}} = \dfrac{36}{1} = 36.$

Instead of multiplying together all the numerators and all the denominators and cancelling down the resulting fraction, we have cancelled as between any numerator and any denominator. See if you can follow the cancellations. A little thought will demonstrate the validity of the process.

A caution must, however, be inserted here; cross-cancelling may only be done across a × sign, never across a +, −, or ÷ sign.

Division, when fractions are involved, is performed by 'turning the divisor upside-down' (first changing mixed numbers into improper fractions, of course) and multiplying by the new fraction so formed. Thus:

$$5\dfrac{5}{8} \div 1\dfrac{1}{2} = \dfrac{45}{8} \div \dfrac{3}{2} = \dfrac{\overset{15}{\cancel{45}}}{\underset{4}{\cancel{8}}} \times \dfrac{\overset{1}{\cancel{2}}}{\underset{1}{\cancel{3}}} = \dfrac{15}{4} = 3\dfrac{3}{4}.$$

EXERCISE 5

1. $\dfrac{2}{3} \times \dfrac{7}{8}$; $\dfrac{1}{6} \times \dfrac{4}{7}$; $\dfrac{3}{4} \times \dfrac{8}{9}$; $\dfrac{5}{8} \times \dfrac{16}{25}$; $\dfrac{3}{7} \times \dfrac{14}{27}.$

2. $3\dfrac{1}{2} \times 2\dfrac{1}{4}$; $2\dfrac{1}{5} \times 5\dfrac{1}{2}$; $2\dfrac{5}{8} \times \dfrac{3}{7}$; $6\dfrac{1}{4} \times 3\dfrac{1}{5}$; $3\dfrac{3}{7} \times 4\dfrac{2}{3}.$

3. $1\dfrac{1}{2} \times \dfrac{3}{4} \times 1\dfrac{5}{9}$; $2\dfrac{1}{2} \times 1\dfrac{3}{5} \times 4\dfrac{1}{2}$; $2\dfrac{1}{2} \times 2\dfrac{1}{2} \times 1\dfrac{1}{2}.$

4. $\dfrac{2}{5} \div \dfrac{3}{4}$; $\dfrac{4}{9} \div \dfrac{1}{2}$; $3\dfrac{1}{3} \div 2\dfrac{2}{9}$; $8 \div \dfrac{2}{3}$; $\dfrac{1}{2} \div 50.$

5. $5\dfrac{1}{7} \div 1\dfrac{1}{5}$; $1 \div \dfrac{1}{18}$; $7\dfrac{1}{2} \div 45$; $10\dfrac{1}{2} \div 2\dfrac{5}{8}$; $100 \div 37\dfrac{1}{2}.$

ADDITIONAL NOTES (FOR THOSE ESPECIALLY INTERESTED)

(*a*) "*Many a mickle makes a muckle.*" Consider the series of fractions:

$$1 + \tfrac{1}{2} + \tfrac{1}{3} + \tfrac{1}{4} + \tfrac{1}{5} + \tfrac{1}{6} + \tfrac{1}{7} + \tfrac{1}{8} + \ldots,$$

the train of dots indicating that the 'run' of fractions is to be continued *ad lib*. The reader will surely be able to continue this series at will.

It would appear, at first thought, that when (say) the hundredth term——$\tfrac{1}{100}$——was reached the addition of subsequent terms in the 'run' would make a hardly appreciable difference to the sum of the series. Yet this is not so. For consider the series grouped as follows:[1]

$$1 + \tfrac{1}{2} + (\tfrac{1}{3} + \tfrac{1}{4}) + (\tfrac{1}{5} + \tfrac{1}{6} + \tfrac{1}{7} + \tfrac{1}{8})$$
$$+ (\tfrac{1}{9} + \ldots + \tfrac{1}{16}) + \ldots.$$

The brackets, or parentheses, have been put in to isolate the two terms finishing at the fourth, the next four terms finishing at the eighth, the next eight terms finishing at the sixteenth, and so on.

Now $\tfrac{1}{3}$ is greater than $\tfrac{1}{4}$; so $\tfrac{1}{3} + \tfrac{1}{4}$ is greater than $\tfrac{1}{4} + \tfrac{1}{4}$, *i.e.*, than $\tfrac{2}{4}$, *i.e.*, than $\tfrac{1}{2}$.

And $\tfrac{1}{5}, \tfrac{1}{6}, \tfrac{1}{7}$ are each greater than $\tfrac{1}{8}$; so $\tfrac{1}{5} + \tfrac{1}{6} + \tfrac{1}{7} + \tfrac{1}{8}$ is greater than $\tfrac{4}{8}$, *i.e.*, than $\tfrac{1}{2}$.

Similarly, the next eight fractions are together greater than $\tfrac{8}{16}$, *i.e.*, than $\tfrac{1}{2}$; and so on.

Thus the sum of our series is more than

$$1 + \tfrac{1}{2} + \tfrac{1}{2} + \tfrac{1}{2} + \tfrac{1}{2} + \ldots.$$

So, by taking a sufficiently large number of terms we can show that the sum has no limit—in the language of mathematics, that it tends to infinity. "Many a mickle makes a muckle."

[1] For a demonstration of this 'bracket-property' see p. 211.

(*b*) *When the Even Signs are changed.* Consider the series of fractions:

$$1 - \tfrac{1}{2} + \tfrac{1}{3} - \tfrac{1}{4} + \tfrac{1}{5} - \tfrac{1}{6} + \tfrac{1}{7} - \ldots,$$

the signs of the even fractions in this run being negative, those of the odd fractions positive.

We can write it as:[1]

$$(1 - \tfrac{1}{2}) + (\tfrac{1}{3} - \tfrac{1}{4}) + (\tfrac{1}{5} - \tfrac{1}{6}) + \ldots$$

Now $1 - \tfrac{1}{2} = \tfrac{1}{2}$, $\tfrac{1}{3} - \tfrac{1}{4} = \tfrac{1}{12}$, $\tfrac{1}{5} - \tfrac{1}{6} = \tfrac{1}{30}$, and so on.

So the sum of this series is $\tfrac{1}{2} + \tfrac{1}{12} + \tfrac{1}{30} + \ldots$, and so it is greater than $\tfrac{1}{2}$.

But we can also write it as[1]:

$$1 - (\tfrac{1}{2} - \tfrac{1}{3}) - (\tfrac{1}{4} - \tfrac{1}{5}) - (\tfrac{1}{6} - \tfrac{1}{7}) - \ldots.$$

Here again, $\tfrac{1}{2} - \tfrac{1}{3} = \tfrac{1}{6}$, $\tfrac{1}{4} - \tfrac{1}{5} = \tfrac{1}{20}$, etc.; and so the sum is also $1 - \tfrac{1}{6} - \tfrac{1}{20} - \ldots$. The sum is therefore less than 1.

But we have already seen that the sum is greater than $\tfrac{1}{2}$.

So we have shown that the sum of this series, however many terms of it are taken, lies between $\tfrac{1}{2}$ and 1.

(*c*) *The Study of Zero.* Reference has already been made to one aspect of zero. If I have 10*s*. in my pocket, and my wife asks for, and takes it, I have nothing left in my pocket—no shillings, if you like.

This is not to be confused with the place-holding zero such as we have in 306, except possibly in the sense that the symbol 0 indicates that there are no tens. Compare also the case where the result of multiplying 36 by 10 is shown by 360.

But there is also an approach to zero via fractions. Consider the 'run' of fractions $\tfrac{1}{2}, \tfrac{1}{3}, \tfrac{1}{4}, \ldots \tfrac{1}{10}, \ldots \tfrac{1}{100}, \ldots \tfrac{1}{1000}, \ldots$ and so on. By proceeding sufficiently far

---

[1] For a demonstration of this 'bracket-property' see pp. 211–212.

in this 'run' we can arrive at a fraction which, while it is greater than 0, yet is so small as to be almost negligible in value. We say that the fractions in this run *tend to zero*. For all practical purposes the millionth fraction, shall we say, of the run is much nearer to zero than our limited minds can imagine, or our most delicate instrument can measure in length units; yet it is not zero.

(*d*) *An Aspect of Infinity*. Let us return once more to our 'run' of fractions $\frac{1}{2}, \frac{1}{3}, \frac{1}{4}, \ldots \frac{1}{10}, \ldots \frac{1}{100}, \ldots \frac{1}{1000}, \ldots$ If we divide unity (1) by each of these fractions in turn the answers are: 2, 3, 4, . . . 10, . . . 100, . . . 1000, ... As we reach the smaller fractions, so does our answer mount higher and higher. If we divide 1 by one-millionth the answer is one million. In fact, by taking as the divisor a fraction sufficiently small, we can produce an answer greater than any number which we may previously designate. Speaking generally, we say that as the fraction by which we are dividing tends to zero the result of the division of 1 by the fraction tends to infinity.

The reader will notice how this conception of infinity as something greater than anything which can be previously designated differs from the idea of infinity as used in everyday speech.

CHAPTER V

DECIMALS AND PERCENTAGES

WHAT ARE DECIMALS?

IT is strange that men waited until the sixteenth century to complete our present Hindu-Arabic system of writing down numbers by extending it backward, as it were, to include fractions. The idea was first developed in southern India at the beginning of the fourth century A.D. Then it spread to the Arabs, and from them, largely over trade-routes, to medieval Europe, where Leonardo of Pisa (about 1500) was the first to introduce it, though not quite in the form which we know to-day. We know that Stevinus, a Dutchman born at Bruges in Belgium in 1548, used decimal fractions, though he did not write them in quite the same way as we do.

The decimal system did not start with the decimal point. It was already inherent in the way in which whole numbers were written down. Most of us can dimly remember in our early childhood writing down addition and subtraction sums with h above the hundreds column, t above the tens column, and u above the units column. They were put there as reminders of the value of each figure in its proper column.

This is of the essence of the decimal system. A digit (or figure) in any particular column has one-tenth the value which it would have in the column next to the left. Thus 5714 means 5 thousands, 7 hundreds, 1 ten, and 4 units. This process may be continued to the right, so long as the units place is marked in some definite way. This is done by means of the decimal point. Thus the decimal point tells us where the whole numbers finish and the fractional part begins. It is the 'full stop' to whole numbers, but it is written a little higher than is usual with a full stop.

The decimal point was not generally used until late in the eighteenth century. The story is told of Lord Randolph Churchill, father of Mr Winston Churchill, that when he was Chancellor of the Exchequer a sheet of figures was placed before him. Noticing the decimal points, he inquired, "And what are those ... dots?" The story is almost certainly apocryphal.

It is interesting to note that in America the decimal point is written in the usual 'full stop' position—

A PAGE FROM "ARITHMETICK VULGAR AND DECIMAL," BY ISAAC GREENWOOD (1729)

e.g., 4·36 is written 4.36; on the Continent a comma is used for the decimal point.

We read 5714·365 as five seven one four point three six five. In terms of fractions this is:

$$5714 + \frac{3}{10} + \frac{6}{100} + \frac{5}{1000}$$

—that is, $5714\frac{365}{1000}$, on adding the fractions.

If one of the figures is missing we put in a 0 as a place-holder. Thus 2·305 is read two point three 0 five. The 0 indicates that there are no hundredths, and so

$$2\cdot305 = 2 + \tfrac{3}{10} + \tfrac{5}{1000}$$
$$= 2\tfrac{305}{1000} = 2\tfrac{61}{200}, \text{ on cancelling down.}$$

It will thus be seen that a decimal—or decimal fraction, to give it its full name—is often a most convenient way of writing down a fraction which in its 'vulgar' form may look rather ungainly. Moreover, decimals often can give us a better idea of the size of a number. For instance, we can tell at a glance that 2·305 lies between 2·3 and 2·4 (or between $2\tfrac{3}{10}$ and $2\tfrac{4}{10}$); this gives us a quicker appreciation of its value than $2\tfrac{61}{200}$ does. This point is worthy of study, and we shall return to it later.

ADDITION AND SUBTRACTION

From the nature of the formation of decimals it is clear that we can add and subtract decimals in the same way as whole numbers, *provided the numbers to be added or subtracted are written down so that the decimal points come directly underneath one another*; for example:

Addition.	14·375	*Subtraction.*	37·64
	2·478		18·95
	39·67		18·69
	56·523		

One or two cases occur which are not covered by the ordinary methods of addition and subtraction as applied to whole numbers. Study the following examples:

Addition.   3·34   ·0125
              7·56   ·0375
            10·9$\emptyset$ = 10·9.   ·05$\emptyset\emptyset$ = ·05.

We strike out the last 0's, in the first case because there

are no hundredths, in the second case because there are no thousandths or ten-thousandths.

Subtraction. 26·847   26·847
              12·93   12·930
              ————   ————
                             13·917

As there is no figure under the 7, we can put a 0 there (no thousandths, if you like) and subtract as for whole numbers.

Subtraction. 92·4   92·40
              25·86   25·86
              ————   ————
                             66·54

There is nothing from which to subtract the 6 (no hundredths, if you like), so we put a 0 after the 4 (92·40 is the same as 92·4), and carry on as for whole numbers.

Of course, after a little practice, it is not necessary to write down the 0's as has been done above.

EXERCISE 6

	(a)	(b)	(c)	(d)
1. Find the sums:	6·23	6·5	10·04	23·25
	4·79	7·29	14·7	18
	3·06	3·05	6·96	25·75
2. Find the sums:	4·435	4·034	30·451	11·375
	5·627	5·002	17·236	12
	1·84	8·77	11·927	13·125
	9·3	5·688	9·236	13·5
3. Subtract:	8·29	16·75	·0125	0·54
	4·37	8·5	·006	·49
4. Subtract:	14·6	23·5	100	10
	2·25	9·75	83·5	4·25

5. A breakfast cereal is advertised as containing in every 100 parts:

Wheat bran	82·5	parts.
Sugar	12·75	parts.
Salt	3·25	parts.
Malt flavouring	1·5	parts.

Check that the parts add up to 100.

6. In a season before the War Hammond's batting average was 63·56; Hardstaff's was 54·58. What was the difference in their averages?

MULTIPLICATION AND DIVISION

Multiplication and division by whole numbers of numbers involving decimals are done as for whole numbers, except that when the decimal point is reached at any stage it is at once put down in the answer, thus:

$$23·45 \times 8$$
$$8$$
$$\overline{187·6\emptyset} = 187·6.$$

$$7)\overline{438·2} \div 7$$
$$\overline{62·6}$$

When we divide by 5 and certain other numbers it may be necessary to introduce a zero or zeros at the end of the number which is being divided in order to bring the division to an exact conclusion. Thus when we divide 46·2 by 5 and 37·45 by 8 we have:

$$5)\overline{46·20} \qquad and \qquad 8)\overline{37·45000}$$
$$\overline{9·24} \qquad\qquad\qquad \overline{4·68125}$$

We shall discuss 'long' multiplication and 'long' division where decimals are involved at a later stage.

EXERCISE 7

1. 16·25 × 4; 8·75 × 8; 43·2 × 5; 29·6 × 7; 42·9 × 12; 7·345 × 4; 2·0036 × 9.

2. 14·8 ÷ 4; 17·1 ÷ 9; 28·2 ÷ 5; 7·6 ÷ 8; 3·876 ÷ 12; 1·764 ÷ 6; ·025 ÷ 4.

FRACTIONS AND DECIMALS

We saw at p. 67 that decimals are really a convenient way of writing down fractions whose denominators are 10, 100, 1000, etc., the denominators being left out, the position of a digit with respect to the decimal point being a sufficient indication of the particular denominator which has been omitted.

Thus $2 \cdot 75 \quad = 2 + \frac{7}{10} + \frac{5}{100}$, or, more conveniently,
$\qquad = 2\frac{75}{100}$, which, on cancelling down,
$\qquad = 2\frac{3}{4}$.

Similarly,
$\quad 9 \cdot 375 \ = 9\frac{375}{1000}$
$\qquad = 9\frac{3}{8}$, on cancelling down.

The old-fashioned rule—and it cannot be beaten—is: to change a decimal fraction to a vulgar fraction write down the decimal fraction, omitting the decimal point; under it place, as denominator, 1 for the decimal point, followed by a 0 for each digit (including zeros) in the decimal fraction. Then cancel down the fraction to its lowest terms. Do you see how it works for the two examples given above?

To change a fraction to a decimal we go right back to the first principles of fractions. At p. 52 it was shown that one of the meanings that can be given to a fraction is 'numerator divided by denominator.' Consider the fraction $\frac{3}{4}$. We can divide the numerator by the denominator by using decimals. Thus:

$$4 \overline{) 3 \cdot 00}$$
$$\cdot 75$$

(4 is not contained even once in 3, so we write down the decimal point and the 3 is 'carried'; then $30 \div 4$, and so on.)

AN AWKWARD POINT

Here we must pause a while, for we are on the brink of another discovery. Parodying some well-known words, we shall see that "to every decimal a fraction, but not to every fraction a decimal." Let us try to change $\frac{2}{3}$ to a decimal:

$$3\overline{)2\cdot000\ldots}$$
$$\cdot666\ldots$$

We find that we can go on dividing by 3 as long as we like (bringing in place-holding 0's *ad lib.*), and there is evidently no prospect of an exact conclusion to our division. This is written: $\frac{2}{3} = \cdot\dot{6}$, the dot over the 6 indicating that it is repeated unendingly.

Take again $\frac{4}{7}$.

$$7\overline{)4\cdot000000\,|\,00\ldots}$$
$$\cdot571428\,|\,57\ldots$$

Here we are up against the same difficulty, but whereas in the case of $2 \div 3$ the figure 6 is repeated without end in the answer, in the case of $4 \div 7$ the repetition does not commence until six figures have been written down in the answer. It is clear that the same six figures will be repeated unendingly. We write this: $\frac{4}{7} = \cdot\dot{5}7142\dot{8}$, the dots over the 5 and the 8 indicating that the figures from the 5 to the 8 are repeated unendingly.

Decimals of this sort, in which one or more figures keep repeating themselves in exactly the same order, are called *recurring*, or *non-terminating*, decimals. The older text-books all contained instructions on how to change a recurring decimal to a fraction; our business here is to take note of the fact that a 'heavy-looking' decimal such as $\cdot875$ is the same as the simple fraction $\frac{7}{8}$, while a simple fraction like $\frac{2}{3}$ or $\frac{4}{7}$ cannot be expressed as an exact decimal.

We shall be meeting these apparent contradictions from time to time in our survey of elementary mathematics. For the moment we shall just say that the explanation is that only vulgar fractions with certain denominators can be equivalent to fractions with the special denominators of decimal fractions (10, 100, 1000, etc.).

CUTTING OFF THE TAIL OF A DECIMAL FRACTION

We now come to the question of how the 'tail' of a long decimal fraction is cut short. This is an important practical matter, for in real-life problems the answers do not by any means always 'come out exactly.'

Let us suppose, for the sake of argument, that the answer to a problem comes to 8·76452..., the train of dots indicating that further working would produce more decimal places if these were required.

If great accuracy is not required it might be sufficient to know the answer to the nearest whole number (as, for instance, if the answer were in pence). The question then arises—would 8 or 9 be the better answer to give? The reader will probably say at once—8·7 (for we need not worry about the rest of the 'tail') is nearer to 9 than it is to 8, so 9 is the obvious answer—and he or she would be correct. But what if we are confronted with a number like 8·5? In this case mathematical definition comes to the rescue; it is agreed that 5 shall belong to the 6–9 class and not to the 1–4 class.

So 8·5 to the nearest whole number is taken as 9.

We call this operation *correcting*, or *rounding off*, a number. It is the mathematical way of giving a number in round figures.

Of course, we can correct, or round off, a number to any decimal degree of accuracy.

Thus:

8·76452 corrected to one decimal place is   8·8.

8·76452 corrected to two decimal places is   8·76.

8·76452 corrected to three decimal places is 8·765.

8·76452 corrected to four decimal places is 8·7645.

And in the same way:

25 lb. 9 oz. corrected to the nearest lb. is 26 lb.

£247 9s. 3d. corrected to the nearest £ is £247.

We also speak of numbers corrected, or rounded off, to the nearest ten, hundred, and so on. Thus:

4528 rounded off to the nearest ten is 4530.

4528 rounded off to the nearest hundred is 4500.

Large figures of exports, etc., given in our newspapers are almost invariably rounded off, sometimes to the nearest million. Thus:

£394,375,426 rounded off to the nearest million £ is £394,000,000, or £394m, to use a modern notation.

£394,375,426 rounded off to the nearest £100,000 is £394,400,000.

'SIGNIFICANT' FIGURES

The reasonableness of an answer does not always depend upon the number of decimal places to which it is given or corrected. Consider, for instance: if a mistake of 1 oz. or even of 1 lb. is made in a ton of coal it is not noticeable; but if a mistake of 1 oz. is made in the weight of 1 lb. of tea it would probably be detected at once.

Therefore it is not the *size* of the error which is important, so much as *how great it is in comparison with the total involved.*

Consider the number 1928·3756.

If we replace it by 1928 the error, ·3756, is very small in comparison with 1928; we say that corrected to four *significant* figures 1928·3756 becomes 1928.

(The rule as to increasing or not altering the last figure in the answer is the same as given for 'correcting' at pages 73 and 74.)

For many purposes it would be sufficient to correct it to three significant figures. It would then become 1930 (making use of the 5-and-over principle referred to above). In this case the fractional error is less than $\frac{2}{1928}$; it could not be more than $\frac{5}{1928}$, and for a three-significant-figure correction the greatest possible error is 5 in 1000. Why?

Consider now ·0093047, which is less than 1. Any error which arises through correcting to a given number of significant figures must now be compared with 93047 and not with ·0093047.

Corrected to four significant figures it becomes ·009305.

Corrected to three significant figures it becomes ·00930 (the 0 *must* be inserted here).

Corrected to two significant figures it becomes ·0093.

Thus we have the double rule for counting significant figures:

If the number is greater than 1 *they are counted from the first figure.*

If the number is less than 1 *the counting is started immediately after any* 0's *which directly follow the decimal point.*

Note. A nought which does not directly follow the decimal point *is* a significant figure. *E.g.*, in ·0093047 the first two 0's are not significant figures, the third 0 is a significant figure.

DECIMALS OR FRACTIONS?

The reader has probably seen in the newspapers letters advocating a decimal coinage for this country. That suggested by the Decimal Association, whose main object is currency reform, is based on £1 (as at present) equal to 10 florins (also as at present) of 100 mils—*i.e.*, £1 = 1000 mils instead of 960 farthings. The following new coins are also suggested: double florin, half and quarter florins, ten, five, four, two, and one mils. The three coins of lowest denomination would have the effect of reducing the values of our present 1*d*., ½*d*., and ¼*d*. by one twenty-fifth.

The arguments in favour of such a coinage may briefly be summarized as:

(*a*) time would be saved during education and life;

(*b*) the four fundamental arithmetical operations would be much simplified;

(*c*) it would be easier to compute dividends expressed as percentages; and

(*d*) it would be simpler to convert our currency to foreign moneys.

A great argument against it is that it would mean re-minting all our coinage unless paper money became universal. It goes without saying that unless our weights and measures were at the same time put on a decimal basis utter confusion would result. We have 16 oz. to 1 lb. ($\frac{1}{16} = \cdot0625$), 14 lb. to a stone ($\frac{1}{14} = \cdot0714\ldots$), 12 inches to a foot ($\frac{1}{12} = \cdot0833\ldots$), and so on. There would be headaches for buyers and sellers alike!

THE METRIC SYSTEM

On the Continent, and in scientific work in all countries, the metric system of weights and measures (and money) is in use. The metric system, adopted first in France in 1801, was one of the consequences of the French Revolution.

A very 'modern' note is apparent in the fact that a scientific committee which had been set up to work on the metric system was purged of some of its most valued members on the plea that it should consist only of men "worthy of trust because of their republican virtues and their hatred of kings"!

In this system the metre is taken as the unit of length. It was supposed to be one ten-millionth of the distance between the North Pole and the Equator, measured along the meridian through Paris. But there was a miscalculation, and the standard metre is the length between two marks on a certain platinum-iridium bar, just as we have a standard yard bar in London. A metre is approximately 39·37 in., and so is comparable with our yard.

All lengths in the metric system are based decimally on the metre. The following is the length table in the metric system:

$$\begin{aligned}
10 \text{ millimetres} &= 1 \text{ centimetre.} \\
10 \text{ centimetres} &= 1 \text{ decimetre.} \\
10 \text{ decimetres} &= 1 \text{ metre.} \\
10 \text{ metres} &= 1 \text{ dekametre.} \\
10 \text{ dekametres} &= 1 \text{ hektometre.} \\
10 \text{ hektometres} &= 1 \text{ kilometre.}
\end{aligned}$$

For lengths which we should estimate or measure in miles the comparative measure in the metric system is the kilometre, and 8 kilometres = 5 miles roughly. For lengths such as we should measure in inches and fractions of an inch, centimetres and millimetres are used in the metric system. The basis of comparison is 1 in. = 2·54 centimetres roughly.

The reader should note the prefixes—in particular, *kilo-*, *centi-*, and *milli-*. *Kilo-* (from a Greek word) stands for 1000 times, and *centi-* and *milli-* (from Latin words) stand for one-hundredth and one-thousandth respectively.

The metric unit of weight is the gram; 1 kilogram = 2·2 lb. approximately. The metric unit of capacity is the litre; 1 litre = $1\frac{3}{4}$ pints approximately. The prefixes *kilo-*, *centi-*, *milli-*, etc., are used with the gram (and sometimes with the litre) for larger and smaller multiples of the unit. The gram, of course, is a very small unit; but even milligrams are quite commonly used in everyday weighing in school science laboratories.

Money is similarly divided in many countries. To name but a few: in France there are 100 centimes to a franc, in Italy 100 centesimos to a lira, in Germany 100 pfennigs to a mark, in the U.S.A. and Canada 100 cents to a dollar. One advantage of this is that the introduction to decimals is made easier for the student!

In the U.S.A., while the British length and capacity units are followed, it is interesting to note that the weights system is semi-metric—or perhaps one should say more literal than the British system, for they have 100 lb. to a hundredweight. This gives 2000 lb. to a ton—a short ton, as we should call it.

ANOTHER TYPE OF FRACTION

It is not generally realized that percentages are just another type of fraction. Like decimals, percentages are fractions with a special denominator. The special denominator is 100, but instead of writing down 100 we use the words *per cent*. If this is remembered the truth will be realized of the saying "We talk of and write of percentages, but when it comes to using percentages we turn them into fractions." 'Cent.,' of course, is derived from the Latin word *centum* for 100. The symbol for per cent. (%) goes back to the seventeenth century, when it was written per $\frac{0}{0}$, and possibly even earlier than that.

To express a fraction as a percentage we express the fraction in hundredths, thus:

$$\frac{3}{4} = \frac{3 \times 25}{4 \times 25} = \frac{75}{100},$$

and, leaving out the denominator, we say $\frac{3}{4} = 75\%$.

This is readily seen to be equivalent to multiplying the fraction by 100 and writing 'per cent.' after the answer.

Thus $\frac{1}{8} = \frac{1}{8} \times 100\% = 12\frac{1}{2}\%$, on cancelling down.

With decimals the same method is followed.

Thus $\quad\quad \cdot24 = \cdot24 \times 100\% = 24\%.$

$\quad\quad\quad\quad \cdot486 = \cdot486 \times 100\% = 48\cdot6\%.$

The reverse process is now clear. To change a percentage to a fraction omit the 'per cent.' and replace the 100 in the denominator.

Thus $5\% = \frac{5}{100} = \frac{1}{20}$, on cancelling down.

$22\frac{1}{2}\% = \dfrac{22\frac{1}{2}}{100} = \dfrac{45}{200}$ (multiplying the top and bottom line by 2 to 'clear' the fraction)

$\quad\quad = \dfrac{9}{40}$, on cancelling down.

To change a percentage to a decimal the operation is even easier. Dividing by 100 just means a shift of the decimal point two places to the left.

Thus

$$10\% = \cdot1\emptyset = \cdot1; \quad 256\% = 2\cdot56.$$

Notice that 100 per cent. of anything is the whole of that thing. More than 100 per cent. of anything means more than the whole of it. It is possible to make a profit of more than 100 per cent. when selling anything, but a loss of more than 100 per cent. is scarcely possible unless you pay a person to take away anything which you are anxious to dispose of.

USING PERCENTAGES

Much of the confusion which arises when we are dealing with percentages would be avoided if it were generally realized that there are three basic types of percentage problems. An example follows of one of each type.

(*a*) *Finding a Percentage of Anything*

A salesman is allowed a commission of 4% on his sales. In one week he took orders for £120 of goods. What was his commission?

$$\text{Commission} = 4\% \text{ of } £120 = \frac{\overset{1}{\cancel{4}}}{\underset{25}{\cancel{100}}} \text{ of } £\overset{24}{\cancel{120}} = £\frac{24}{\underset{5}{5}}$$

$$= £4\tfrac{4}{5} = £4 \ 16s. \ 0d.$$

(*b*) *Finding what Percentage one Thing is of Another*

In one week influenza deaths increased from 125 to 135. What was the increase per cent.?

Increase $= 135 - 125 = 10.$

Increase expressed as a fraction $= \tfrac{10}{125}.$

Increase expressed as a percentage

$= \tfrac{10}{125} \times 100 = 8$, on cancelling down.

Note. The original number of deaths is put in the denominator of the fractional increase, not the increased number. Similarly, in the calculation of a percentage error the correct figures are placed in the denominator, not the incorrect ones.

(*c*) *Finding a whole of which a Percentage is given*

57 per cent. of the electorate voted at a by-election. How many electors are there on the register if 11,400 voted at the election?

$$57\% = \tfrac{57}{100}.$$

So $\tfrac{57}{100}$ of the electors number 11,400,
and so the whole number of electors is $11,400 \div \tfrac{57}{100}$

$$= \frac{11,400 \times 100}{57}$$

$= 20,000$ on working out the fraction.

EXERCISE 8

1. Give approximate answers to the following:
   49% of a number is 21; find the number.
   25½% of a number is 30; find the number.
   19⅞% of a sum of money is £2; find the sum of money.
   33% of a sum of money is 6s. 8d.; find the sum of money.

2. A seedsman guarantees that at least 85% of the seeds in a packet will germinate. A packet of sweet-pea seeds contains 36 seeds. What is the least number of plants that can reasonably be expected?

3. About 70% of a pig is meat. How many score must a pig weigh to give about 170 lb. of meat? (Nearest score.)

4. In a Welsh village 95% of the people speak good English, and 84% speak good Welsh. What percentage of the people speak good English and Welsh? (This is a teaser!)

5. A householder insures his house for 72% of its value. If it is valued at £850 for how much does he insure it, and what will the premium come to if it is ½% of the sum insured? (Nearest penny.)

6. What is the discount per cent. if a £10 10s. radio set is sold for £9 16s. cash?

7. A gas company allows a rebate of 5% for prompt payment of accounts. What will be saved by prompt payment of an account of £8 5s.?

8. How must a suit be marked so that when it is sold at a discount of 20% the outfitter still gets £4 10s. for it?

9. The population of a certain village is now 1260. This is 5% more than it was ten years ago. What was the population ten years ago?

10. A landlord expects the rent of one of his houses to be 7½% of what he paid for it. He gave £1200 for the house. Would a rent of 35s. per week be too little or too much, and by how much per year?

CHAPTER VI

THE FUNDAMENTAL ARITHMETICAL OPERATIONS

FOREWORD

IN the first chapter of this book we discussed the beginnings of arithmetic. The merchants of Babylon, the priests of Egypt, the scholar thinkers of Greece, the practical Romans all played their parts. Later the monks of the Middle Ages and the wise men of India and Arabia made their several contributions to its development.

The invention of printing spread abroad the new knowledge, and when, in the sixteenth century, the use of symbols became more common and standardized, the way was open for further advances in mathematics.

But arithmetic, alone in mathematics, stands much where it did some hundreds of years ago. Our knowledge of arithmetic has not increased as has our knowledge of algebra, geometry, and the other mathematical branches. Our methods are perhaps better, but that is largely due to the improvement in the tools which we use.

Some of the books published during the last seventy or eighty years are not nearly so 'modern' in some respects as Robert Recorde's *Grounde of Artes*, printed about 1540. In this book, cast throughout in the form of a dialogue, he made use of many modern 'practical' ideas. He made his introductions to new processes by means of topical illustrations likely to interest the student. He gave many examples; tricky points in the various processes were introduced one by one. And

¶The Grounde of Artes:
teaching the perfecte worke and
practise of Arithmetike, both in whole nū-
bers and fractions; after a moxe easie and
exact soxt, than hitherto hath bene set foxth.

Made by M. ROBERT RECORDE, D. in Physick,
and afterwards augmented by M. IOHN DEE.

And now lately diligently coxrected, & beau-
tified with some newe Rules and necessarie Additions:
And further endowed with a thirde part, of Rules of
Pxactize, abxidged into a bxiefer methode than
hitherto hath bene published: with diuerse
such necessary Rules, as are incident
to the trade of Merchandize.

Whervnto are also added diuers Tables & instructions
that will bxing great pxofite and delight vnto Merchants, Gentlemen, and others, as by the con-
tents of this treatise shal appeare.

By *Iohn Mellis* of *Southwark*, Scholemaster.

Impxinted by I. Harison, and H. Bynneman.
ANNO DOM. 1582.

THE TITLE-PAGE OF THE 1582 EDITION OF "THE
GROUNDE OF ARTES," BY ROBERT RECORDE

he taught the use and value of checks on working. Thus:

> MASTER. But now, can you tell how to prove this addition, . . . and to try whether you have well done or no?
> SCHOLAR. I would I could.

And after finding square roots:

> MASTER. Although I know it to be so, yet for your better exercise, and full persuasion, I would have you try it by square multiplication.
> SCHOLAR. That may I soon do. And so I find it true. As this work here set doth show.

Another idea which many people regard as very up to date, the 'making change' method of addition, first appeared in a French arithmetic book in 1559.

OUR HAPHAZARD BRITISH SYSTEM OF UNITS

When we consider addition and subtraction it must strike us that we in this country are accustomed to working in many different systems of units. On the Continent addition and subtraction depend entirely on the ten, or denary, system. Here we use the ten system for numbers and decimals, a twelve-twenty system for £ s. d., a twelve-three-twentytwo-ten-eight system (or one of its many variations) for lengths, a two-four system for capacity (pints, quarts, gallons), and so on.

EXCHEQUER STANDARD
WINCHESTER BUSHEL
OF HENRY VII

But we have grown up with these units, and with practice they present no more difficulty than does the metric system.

Of course, our very varied system has developed in a haphazard manner. The names of £ s. d. we owe to the

Romans (£ for *libræ*, pounds; *s* for *solidi*, shillings; *d* for *denarii*, pence). Until national standards were finally fixed by law in 1878 many of the units of weights and measures were only of local significance.

Strangely enough, one of the provisions of Magna Carta (1215) was:

> There shall be one measure of wine and one of ale through our whole realm; and one measure of corn, that is to say, the London quarter; and one breadth of dyed cloth . . ., that is to say, two ells within the lists (that is, between the selvages); and it shall be of weights as it is of measures.

Yes, our units are complicated enough; but ask a student of Electricity and Magnetism what he thinks of their units—and they are worked on the metric system!

HOW MAN DID HIS SUMS

Let us start at the very beginning. We, in our addition and subtraction, make use of simple number-facts such as 3 plus 4 is equal to 7, 9 minus 5 is equal to 4, and so on. This is a comparatively modern development. Robert Recorde in his *Grounde of Artes*, from which we have already quoted, says:

> Now that you have learned the common kind of Arithmetic with the pen; you shall see the same Art in Counters: which feat doth not only serve for them that cannot write or read, but also for them that can do both, but have not at sometime their pen or tables ready with them.

Our childish arithmetic started with counting. There is a great deal behind that word; it goes back very far into the social history of mankind. For once upon a time nearly all calculating was done mechanically on an instrument called an abacus.

There were two forms of abacus. One was a table or frame on which counters were fastened by means of grooves, wires, or rods—in one form much like the bead frame with which we played when sitting in our high chairs twenty, thirty, forty, fifty years ago.

The other form—it was used in Mediterranean lands four hundred years before the birth of Christ and was still used in this country three hundred years ago—consisted of loose counters in the form of small disks of metal, bone, glass, or other material, which were placed on or between lines marked upon a table or board.

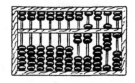
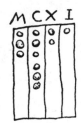

THREE DIFFERENT KINDS OF ABACUS

On the left is the Russian abacus, still used to-day. The middle drawing represents the principle of the Roman abacus. On the right is the Chinese abacus, still used to-day.

It is worthy of note that the abacus is still used all over the world. Up to a few years ago the Russian abacus, the *s'choty*, formed a regular part of the work in the third year in school; in Japan *soraban* calculation, a kind of counter calculation, is a regular branch of school-work; and in China native clerks employed in foreign banks prefer to use the *suan-pan* to Western calculating machines—and they can use it with almost unbelievable speed!

To return to the counting board. As we should expect from a device which has been in use for more than 2000 years, it has left its mark on our language. The very word 'calculate' is derived from the Roman word for pebbles, often used as counters. Business men 'cast their accounts.' Some business houses have their 'counting house,' and shop assistants stand behind the 'counter.' We all 'borrow 1' or 'carry 2' when adding or subtracting numbers—a reference to the way in which a counting board was operated. And the Chancellor of the 'Exchequer' balances the nation's 'accounts' largely by means of 'cheques' drawn by the long-suffering tax-payer. One wonders what derivation the ordinary people of a hundred years hence will attribute to our modern expression 'under the counter.'

It is not our purpose in this book to discuss how an abacus of either form was used; the interested reader will find all he requires on this subject in a good history of mathematics.

AIDS TO SPEED AND ACCURACY IN ADDITION AND SUBTRACTION

The average man or woman approaches a complicated addition or subtraction with something akin to trepidation. Over and over again the operation is performed, much time and temper are wasted, and sometimes the same answer is obtained twice. What can be done about it? Here are some well-tried aids to speed and accuracy.

1. In adding, say, 4, 8, 5, 7, get accustomed to saying 4, 12, 17, 24, and not 4 and 8 equals 12, and 5 equals 17, and 7 equals 24. Not only does this save time, but the very fact that time is saved means that the mind does not tire, and therefore become less alert, when dealing with long columns of figures.

2. Time may also be saved by looking ahead to convenient combinations of numbers. For example, in adding 6, 8, 4, 12, 5, if the order is altered (mentally) to 6, 4, 8, 12, 5, the addition simplifies itself to 10, 30, 35.

3. Check column additions by doing the first addition upward, the second downward. This is a good habit to

form, and should be strictly adhered to. In the case of adding across the page, first add from left to right; check by adding from right to left.

4. In the addition of two rows, further check by subtracting either row from the sum.

5. In subtraction check by adding the difference to the quantity subtracted.

6. Those whose work entails much addition should practise adding two columns simultaneously: thus, adding 25, 17, 32, 46, say 25, 42, 74, 120. Practice makes perfect.

7. Another time and trouble saver:

Suppose we have $57 + 19$. This is more easily thought of as $57 + 20 - 1$, or $77 - 1$—that is, 76. Similarly, $35 + 97$ is $35 + 100 - 3$—that is, $135 - 3$, or 132; and $48 + 75$ is $48 + 70 + 5$, or $118 + 5$—that is, 123.

The same idea is readily applied to £ s. d. Thus, $3s. 7d. + 11d. = 4s. 7d. - 1d. = 4s. 6d.$; $14s. 8d. + 9d. = 15s. 8d. - 3d. = 15s. 5d.$

8. Here is something which should be a help to those who want to improve their mental addition.

When we do an addition sum in writing we always start with the right-hand column; if we do it in our heads experience shows some people find it better to start from the left. It may be that our natural mode of progress is from left to right, not from right to left. Thus, if we add 237 to 489 in writing we shall have

$$
\begin{array}{r}
237 \\
489 \\
\hline
726
\end{array}
$$

Mentally: (hundreds) 6

           (tens)   11   making 71

           (units)   16   making 726

This process is difficult to explain in words, but the reader may find it easier to keep the first figure, then the first two figures, and so on, in his head while he is working on the next column, than if he starts in the usual way with the right-hand column and tries to keep the last figure, then the last two figures, etc., in his head.

9. A final check: see that the answer is not unreasonable—too large or too small.

EXERCISE 9

1. Add:

1234	5689	3928	6758
2143	2136	72	429
4312	4625	405	7328
4321	1584	3500	4511
			3922
			7184

2. Add crossways and downward, checking by adding the cross-sums and the down-sums:

Cross-sums

23	17	15	77	14
46	28	87	22	26
87	34	25	33	38
59	82	43	44	50
28	19	66	11	62

Down sums   Check here

3. Find the difference, checking the answers:

68	75	134	286	400	5062	10000
39	57	17	77	23	3184	6543

4. Add:

£	s.	d.	£	s.	d.	£	s.	d.	£	s.	d.
3	2	5	13	17	$2\frac{1}{2}$	28	18	6	234	15	7
4	8	$6\frac{1}{2}$	15	2	9	39	14	2	1796	7	9
	3	9	24	3	$10\frac{1}{2}$	16	10	$8\frac{1}{2}$	542	4	3
5	2	10		14	6	28	4	$7\frac{1}{2}$	827	19	5
6	8	$5\frac{1}{2}$	5	2	8	34	17	$9\frac{1}{2}$	560	18	8
9	17	8				25	0	0	432	10	11

5. Find the difference, checking the answers:

£	s.	d.	£	s.	d.	£	s.	d.	£	s.	d.
42	3	11	2	11	7	4	13	7	235	16	8
25	5	6	1	6	8	2	15	5½	194	18	10

6. For practice in mental addition use the figures given in questions 3 and 4 above, adding as well as subtracting.

MULTIPLICATION AND DIVISION

Just as in addition and subtraction we subconsciously make use of simple number-combinations like 4 plus 3 is equal to 7, so in multiplication and division we constantly hark back to the multiplication tables which were once so vigorously thumped into us.

But this was not always done. Up to a few hundred years ago even what tables there were finished at the 'five times' (our hands have always been responsible for many of our shortcomings); and so when anyone wished to multiply, say, 8 by 6 he did it like this:

Subtract each of the numbers from 10 and write them and the subtracted results as follows:

$$8 \quad 2$$
$$6 \quad 4$$

For the units digit of the product multiply the 2 by the 4, giving 8; for the tens digit subtract 4 from 8 (or 2 from 6, the result in each case always being the same), giving 4.

$$8 \times 6 = 48.$$

Another way of avoiding learning the multiplication table beyond five times five was the practice known as 'finger multiplication.' This is how it is applied to 8×6:

Hold out both hands with fingers (this includes thumbs) outstretched.

Turn down 8 − 5, or 3, fingers on one hand.

Turn down 6 − 5, or one, finger on the other hand.

On the first hand there are two fingers still standing; on the other hand 4 fingers still standing.

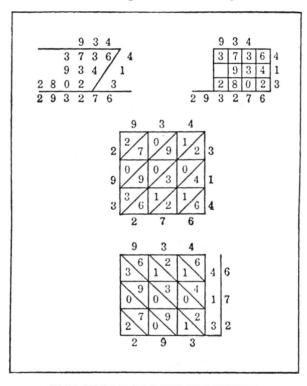

FOUR METHODS OF MULTIPLICATION

Adapted from the *Treviso Arithmetic* (1478). Modern numerals have been used instead of those in the original.

The tens digit of the product is the total number of fingers turned down on both hands—that is, 4; the units digit of the product is obtained by multiplying together the numbers of those still standing, 2 × 4 = 8.

$$8 \times 6 = 48.$$

The reader who is sufficiently well up in algebra may like to prove now, or at a later stage, that the method depends upon:

$$ab = 10\left\{(a - 5) + (b - 5)\right\} + (10 - a)(10 - b).$$

Rather laborious, you will say, but the most common alternative was repeated additions (for multiplication) or subtractions (for division) on the abacus. It is on record that Pepys the diarist, in spite of the best education of his day, had to get up in the summer at 4 A.M. to hammer in the multiplication tables not included in his learning, so that he could do the Admiralty accounts.

From this the reader will appreciate the absolute necessity of mastering mechanical operations, even if it entails the working out of many examples. The feeling that one is gaining in power and skill more than compensates the time expended.

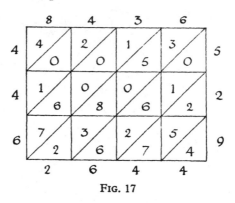

FIG. 17

There were many methods of 'long' multiplication other than those in general use nowadays. Perhaps the reader may care to while away a few minutes finding out how the 'lattice,' or 'Chinese,' method of multiplication is performed. Only one hint is given; the final addition, done vertically in ordinary multiplication, is here done on a slope.

Division by the methods which we use now came in about five hundred years ago, but for nearly three hundred years after that the 'scratch,' or 'galley,' method here illustrated for the division of 65,284 by 594 remained popular. The derivation of its name is obvious; the reasons for its popularity are not.

1̸5
5̸3̸3
1̸6̸8̸7̸8
6̸5̸2̸8̸4̸(109
5̸9̸4̸4̸4̸
5̸9̸9̸
5̸

THE 'SCRATCH'
METHOD OF
DIVISION

MODERN PRACTICE IN MULTIPLICATION AND DIVISION

Much discussion has raged, and still rages, over the best way of setting down multiplication; but current opinion tends to favour the method in which a start is made with the left-hand digit of the multiplier (not the right-hand digit, as in former days), the first figure of each product being set down under the multiplying figure at each stage. Study this example, and notice how these points are displayed in it:

$$
\begin{array}{r}
4738 \\
2065 \\
\hline
9476 \\
28428 \\
23690 \\
\hline
9783970 \\
\hline
\end{array}
$$

To multiply when decimals are involved, perhaps the best way is still to multiply together the two numbers as if there were no decimal point present; then fix the position of the decimal point in the product by making the number of decimal places in the product equal to the sum of the numbers of decimal places in the decimals which are being multiplied.

In division modern practice is to place the first figure in the quotient above the right-hand digit of the first dividend (not to the right, as in the old-fashioned way). In case a reminder is necessary—the *dividend* is the number which is being divided. For example:

```
                           241   (Quotient)
        (Divisor)     365)88310
                           730

                          1531
                          1460

                           710
                           365

                           345   (Remainder)
```

Here the first dividend is 883, so we put the 2 (883 ÷ 365 = 2 + ...) over the 3, then carry on in the usual way.

In dividing where decimals are involved the old rule is perhaps still the best: make the divisor a whole number by moving the decimal point as many places to the right as is necessary; move the decimal point in the dividend an equal number of places to the right; and when the decimal point is reached in the division put it down at once in the quotient and carry on. The reason for this is that if we regard the operation as numerator divided by denominator we have made the denominator a whole number by moving the decimal point; to obtain the equivalent fraction we have to do the same to the numerator. But, as we shall see later (Chap. XX), these mechanical processes of multiplication and division are best done by logarithms.

EXERCISE 10

1. Find the products:

32	57	36	37	67
24	38	36	27	19

2. 312×24; 505×61; 4372×75; 126×140; 342×204.

3. $3\cdot8 \times 4\cdot2$; $5\cdot6 \times 3\cdot5$; $3\cdot215 \times \cdot48$; $\cdot005 \times \cdot05$.

4. Divide: $16\overline{)384}$; $25\overline{)625}$; $31\overline{)837}$; $365\overline{)45,067}$; $144\overline{)1,000,000}$

5. (With decimals.) $1760 \div 128$; $25\cdot46 \div 6\cdot7$; $38\cdot7 \div 4\cdot5$; $3\cdot857 \div 53\cdot2$; $\cdot192 \div 62\cdot5$.

MORE AIDS TO SPEED AND ACCURACY

(a) *Multiplication Checks*

1. It is usual to multiply the number with the larger number of digits by that with the smaller number of digits. If time permits, check by altering the order of the multiplication: thus, if the product of 457 and 18 is required, after multiplying 457 by 18 check by multiplying 18 by 457.

2. A useful rough check which obviates in particular an error in the positioning of the decimal point is done as follows. Take as an example $38\cdot7 \times 14\cdot3$. The product is greater than 35×10, or 350, and it is less than 40×20, or 800. So the product lies between 350 and 800. Of course, the margin of error is still large, but not small enough for the decimal point to slip through. Of course, the margin can be made smaller still.

3. Checks which definitely tell us whether we are wrong (but not necessarily that we are therefore right) are often useful. The fact that multiplication of any number by an even number must give an even number is obvious; the multiplication of two even numbers must give a number whose last two figures form a number exactly divisible by four. So also a number multiplied by a number ending in a five must give a result ending in a five or a nought.

But a check which gives confidence is the test for divisibility by three or nine. If either the number which you are multiplying or the number with which you are multiplying is divisible by three or nine (for test, see p. 36) then the answer must satisfy the same test.

4. Lastly, the answer must be reasonable—not obviously too large or too small.

The reader who is interested will, no doubt, from these hints think of further checks for himself.

EXERCISE 11

Several of the following products are obviously incorrect. Find out which they are without performing the multiplications:

$22 \times 24 = 528$; $19 \times 19 = 363$; $25 \times 15 = 370$; $26 \times 22 = 562$; $45 \times 35 = 1585$; $25 \times 40 = 1000$; $30 \times 24 = 7200$; $4 \cdot 8 \times 3 \cdot 5 = 168$.

(b) *Division Checks*

1. Clearly the safest check of all is to multiply the quotient by the divisor, adding in the remainder, if any.

2. Especially when decimals are involved it is wise, as for multiplication, to find between what limits an answer should be by performing a simplified division.

Consider $3189 \cdot 6 \div 42 \cdot 75$. The answer must be less than $\frac{3200}{40}$, or 80 (for we have made the numerator larger and the denominator smaller, thus increasing the value of the fraction). Reversing the argument, the answer must be greater than $\frac{3000}{50}$, or 60. So an answer which does not lie between 60 and 80 would be in error. Of course we could narrow the range by taking, say, $\frac{3000}{45}$, or $66 \cdot 6$, as our lower limit; but the important thing is to take numbers for the checks which can be cancelled down easily.

3. Here again, the answer must be reasonable—not obviously too large or too small.

(c) Short Cuts in Multiplication

1. To multiply by 10 add on a 0; to multiply by 100 add on two 0's, etc. For example:

$$734 \times 10 = 7340; \quad 251 \times 100 = 25,100; \text{ etc.}$$

This is a particular case for whole numbers of a general rule applying also to decimals; to multiply by 10, 100, ... move the decimal point 1, 2, ... places to the right (filling in with place-holding zeros if necessary). For example:

$$2\cdot35 \times 10 = 23\cdot5; \quad 47\cdot8 \times 1000 = 47,800.$$

2. To multiply by 25 add on two 0's and divide the result by 4. This, of course, is equivalent to multiplying by 100 and dividing by 4, and $100 \div 4$ is equal to 25. For example:

$$37 \times 25 = 3700 \div 4 = 925; \quad \cdot46 \times 25 = 46 \div 4 = 11\cdot5.$$

3. To multiply by 125 add on three 0's and divide the result by 8. The proof is left to the reader. For example:

$$61 \times 125 = 61,000 \div 8 = 7625.$$

4. To multiply by 9 add on a 0 and subtract from the result the number which is being multiplied. The work is best written down as follows:

$$
\begin{aligned}
785 \times 9 = \quad & 7850 \\
- \quad & 785 \\
\hline
= \quad & 7065
\end{aligned}
$$

5. To multiply by 99 add on two 0's and subtract the number which is being multiplied. For example:

$$
\begin{aligned}
369 \times 99 = \quad & 36,900 \\
- \quad & 369 \\
\hline
= \quad & 36,531
\end{aligned}
$$

This rule may be extended as follows:

$$
\begin{aligned}
437 \times 96 = \quad & 43,700 \\
- \quad & 1,748 \quad (4 \text{ times } 437) \\
\hline
= \quad & 41,952
\end{aligned}
$$

G

97

A few examples are now given, more as reminders, of applications to £ *s. d.*

6. To multiply by 12 (or to find the cost per dozen) call pence shillings and parts of a penny the same parts of a shilling. For example:

$$8\tfrac{1}{2}d. \times 12 = 8s.\ 6d.;$$
$$1s.\ 7\tfrac{3}{4}d. \times 12 = 19\tfrac{3}{4}d. \times 12 = 19s.\ 9d.$$

7. To multiply by 20 (or to find the cost per score) call shillings £'s and parts of a shilling the same parts of a £. For example:

$$5s.\ 3d. \times 20 = £5\ 5s.; \qquad 14s.\ 9d. \times 20 = £14\ 15s.$$

The reader is invited to think out similar rules for himself: thus lb. × 112, pence × 240.

EXERCISE 12

1. 87·5 × 10; 43·4 × 100; ·37 × 10; ·0035 × 1000; 3·735 × 100.

2. 42 × 25; 65 × 25; 4·3 × 25; 87 × 125; 5·6 × 125.

3. 287 × 9; 342 × 9; 46 × 99; 87 × 99; 28 × 97; 476 × 999.

4. $7\tfrac{1}{2}d.$ × 12; 2s. $5\tfrac{1}{2}d.$ × 12; 3s. $10\tfrac{1}{4}d.$ × 12; 4s. 3d. × 20; 15s. 6d. × 20; 18s. 9d. × 20; 2 lb. 8 oz. × 112; 3 cwt. 28 lb. × 20; $7\tfrac{3}{4}d.$ × 240; $4\tfrac{1}{4}d.$ × 120.

(d) Short Cuts in Division

1. Division by 10, 100, ... is the reverse of the multiplication process; so to divide by 10, 100, ... move the decimal point 1, 2, ... places to the left. For example:

48·7 ÷ 10 = 4·87; ·025 ÷ 100 = ·00025 (here we have to fill in with two place-holding zeros).

2. To divide by 25 multiply by 4 and divide by 100; to divide by 125 multiply by 8 and divide by 1000. The proof is left to be thought out by the reader. For example:

675 ÷ 25 = 675 × 4 ÷ 100 = 2700 ÷ 100 = 27.

And then muſt I look how often I may find the laſt figure of the diviſor (that is 4) in 13, which I may doe 3 times, therefore do I ſay, 3 times 4 is 12, which I take out of 13, and there remaineth 1. Then do I make at the right hand of my ſummes a crooked line, and write before it my quotient 3, and I cancell 13 and 4, and over the 3 I ſet the 1 that remaineth: and then the figures ſtand thus.

```
      1
136280 (3
452
```

Then I multiply the ſame quotient into every figure of the diviſor, and withdraw the ſumme that amounteth out of the numbers over them: as firſt I ſay, 3 times 5 make 15, which I take from 16, and there reſteth 1: I cancell therefore 16 and 5, and write over the 6 that 1 that remaineth, thus.

```
   1
 136280 (3
 452
```

Then doe I ſay likewiſe, 3 times 2 make 6, which I take out of 12, and there reſteth 6; therefore I cancell the 12 and the 2 over, and then I write the 6 that remaineth, thus.

```
  16
 136280 (3
 452
```

Then ſhould I ſet forward the diviſor into the next place toward the right hand thus.

```
  16
 136280 (3
 4512
 45
```

Maſter. But you may ſee that over the 4 is no figure, therefore I muſt ſet the diviſor yet forwarder by another place.

And mark, Whenſoever it chanceth ſo that you ſhould ſet forward the diviſor, and that it cannot ſtand here, becauſe there is no number over the

G 3

The ſcholar, prompted by the maſter, explains the early ſteps in the diviſion of 136,280 by 452.

$2480 \div 125 = 2480 \times 8 \div 1000 = 19,840 \div 1000 = 19\cdot84$.

3. Reversed rules for dividing money by 12 or 20 may be deduced from those given for multiplication at p. 98. For example:

17s. 9d. $\div$ 12 = $17\frac{3}{4}d$. = 1s. $5\frac{3}{4}d$. If coal is £4 15s. per ton it is $4\frac{3}{4}s$., or 4s. 9d., per cwt.

4. For those who are not too fond of long division the following rather amusing method of division by multiplication is commended. Unfortunately it only works if the divisor has many factors 2 or 5. The object is, by successive multiplications of numerator and denominator by the same numbers, to change the denominator to a single digit followed by zeros—then we use the decimal point and short division. For example:

$$5837 \div 64 = \frac{5837}{64}$$

$$= \frac{29,185}{320} \text{(multiply by 5)}$$

$$= \frac{145,925}{1600} \text{(multiply by 5)}$$

$$= \frac{729,625}{8000} \text{(multiply by 5)}$$

$$= \frac{729\cdot625}{8} = 91\cdot203125.$$

Incidentally, by carrying on with the multiplication by five we can change the denominator to 1,000,000.

EXERCISE 13

1. $3\cdot87 \div 10$; $439 \div 100$; $24\cdot5 \div 1000$; $\cdot375 \div 10$; $8\cdot6 \div 100$.

2. $420 \div 25$; $3456 \div 25$; $7129 \div 125$; $45 \div 125$.

3. 14s. 9d. $\div$ 12; 8s. 3d. $\div$ 12; £2 15s. $\div$ 20; £14 13s. 4d. $\div$ 20; £10 $\div$ 240; £5 5s. $\div$ 120; $15\frac{1}{2}$ cwt. $\div$ 112.

4. (Division by multiplication.) $325 \div 16$; $430 \div 32$; $1679 \div 128$.

CHAPTER VII

APPLIED ARITHMETIC

Now that we have revised the fundamental processes of arithmetic there remains the task of applying them to the arithmetic of daily needs. Let this be remarked at the outset: when the so-called 'four rules'—addition, subtraction, multiplication, and division—are mastered, with their extensions from whole numbers to fractions and decimals, the fundamental processes of arithmetic are learnt. Difficulties that may arise in their application will be mostly matters of comprehension, not of manipulation.

THE IDEA OF PROPORTION

Which of the following statements are true?

(*a*) An aeroplane has a cruising speed of 180 m.p.h.; it will therefore at this speed cover 360 miles in 2 hours and 540 miles in 3 hours.

(*b*) Henry VIII had six wives; therefore Henry IV had three wives.

(*c*) Two pictures which measured 24 in. by 24 in. and 20 in. by 16 in. have the same shape.

(*d*) 5 miles = roughly 8 kilometres. Therefore 80 kilometres = roughly 50 miles.

(*e*) In two hours I caught eight fish; therefore in eight hours I should have caught thirty-two fish.

(*f*) Rations for four people for sixteen days should be sufficient for eight people for eight days.

(*g*) When I was eighteen I weighed 10 stone, so when I am fifty-four I shall weigh 30 stone.

(*h*) An aeroplane with one motor developing 1200 h.p. has a maximum speed of 240 m.p.h. so an aeroplane with four motors, each of 1200 h.p., should have a maximum speed of 960 m.p.h.

(*i*) If three barking dogs keep twelve people awake, then four barking dogs will keep sixteen people awake.

It will not be surprising if you had all or very nearly all the answers correct.

Now why were you able to pick out the 'wrong 'uns'? Because you used your common sense. For instance, it is ridiculous to suppose that a person's weight at the age of fifty-four years is three times that person's weight at eighteen. On the other hand, if the same average speed is kept up, a motorist may expect to travel three times as far in three hours as he does in one hour. And, contrariwise, if a motorist trebles his average speed over a given distance he may expect to cover that distance in a third of the time.

DIRECT AND INVERSE PROPORTION

In the last two examples we have exemplified two very important common-sense principles—those of *direct* proportion and *inverse* proportion. If two quantities are connected in direct proportion doubling one of them means that the other is doubled, and multiplying one of them by, say, $\frac{9}{4}$ means that the other will also be multiplied by $\frac{9}{4}$. If two quantities are connected in inverse proportion doubling one of them means that the other is halved, while multiplying one of them by, say, $\frac{9}{4}$ means that the other is multiplied by $\frac{4}{9}$.

Two quantities which are directly proportional increase or decrease together; if two quantities are inversely proportional then when one increases the other decreases, and *vice versa*.

Of course, there are other kinds of proportion. The

102

areas of squares are proportional to the squares of their sides; the volumes of cubes are proportional to the cubes of their edges. And Newton's law of universal gravitation says that the attraction between two bodies is inversely proportional to the square of the distance between them and directly proportional to the product of their masses (and to this we owe the motion of falling bodies and the facts that the moon moves round the earth and the planets round the sun).

METHOD OF UNITY

Problems involving direct and inverse proportion are readily solved by the *unitary method*, or the *method of unity*. One example of each kind is given to refresh the memory of the reader.

(*a*) Sixteen complete turns of a screwdriver just drive a screw through a $1\frac{1}{2}$-in. plank. How many turns will be required to drive the screw $\frac{3}{8}$ in. into the plank?

To drive the screw $1\frac{1}{2}$ in. requires 16 turns.

To drive the screw 1 in. requires $\dfrac{16}{1\frac{1}{2}}$ turns.

To drive the screw $\frac{3}{8}$ in. requires $\dfrac{16}{1\frac{1}{2}} \times \frac{3}{8}$ turns

$$= \frac{16}{\frac{3}{2}} \times \frac{3}{8} = \frac{16}{1} \times \frac{2}{3} \times \frac{3}{8} = 4 \text{ turns.}$$

Notes. (i) In the first line of the solution we place what we want to find (here the number of turns) last.

(ii) In the second line we have the justification for the name of the method—unitary method. From $1\frac{1}{2}$ in. we come (by division) to 1 in.

(iii) In the third line we introduce $\frac{3}{8}$ in., the distance through which the screw is to be driven; and here we multiply.

(*b*) An accumulator lasts for 8 days without re-charging if a wireless set is used on the average $3\frac{3}{4}$ hours a day. How many days will it last if the set is used an average of only $2\frac{1}{2}$ hours a day?

Used $3\frac{3}{4}$ hours a day the accumulator lasts 8 days.

Used 1 hour a day the accumulator lasts $8 \times 3\frac{3}{4}$ days.

Used $2\frac{1}{2}$ hours a day the accumulator lasts $\dfrac{8 \times 3\frac{3}{4}}{2\frac{1}{2}}$ days

$$= 8 \times \frac{15}{4} \div \frac{5}{2} = \frac{\overset{2}{\cancel{8}}}{1} \times \frac{\overset{3}{\cancel{15}}}{\underset{1}{\cancel{4}}} \times \frac{2}{\underset{1}{\cancel{5}}} \text{ days}$$

$= 12$ days.

Notes. (i) Here again, in the first line of the solution, we place what we want to find—the number of days—last.

(ii) In the second line, the 'unity' line, common sense tells us that the accumulator lasts longer, so we multiply by $3\frac{3}{4}$.

(iii) And so for the longer time, $2\frac{1}{2}$ hours a day, in the third line, we have to divide by $2\frac{1}{2}$.

EXERCISE 14

First decide whether an answer is possible.

1. If a row 32 ft. long takes 48 potatoes how many potatoes will be required for a row 40 ft. long?

2. A car 2 years old is worth £350. What will it be worth when it is 4 years old?

3. Potatoes are marked in a greengrocer's shop at 4 lb. for 5*d*. What will 14 lb. cost?

4. A farmer can plough his land with four tractors in 12 days.

```
              100----6----75
           A  12              y        B |
      l.      l.      l.            then fay
  If 100      6      75     m.  l.  s.   m.
             75           If 12---4--10----9
                               20
             30
             42              90 fhillings
            ------ l. s.      12
      1|00) 4|50 (4·10        ------
            4                 180
                              90
      ------               ------
  Remains (50)              1080 pence
  Multiply 20                9
                            ------    12) 2|0) l.  s.  d.
      1|00) 1c|00 (10 s.   12) 9720 (810 (6|7 (3--7---6

            l.  s.          96      72
   Facit  4---10
                            12      9)
                            12      84
                           ------  ------
                            (0)    (6) pence

                              l.   s.   d.
                      Facit  3---7---6
```

So that by the foregoing Operation I conclude that *if* 100 *l. in* 12 *months gain* 6 *l.* Intereft, 75 *l.* will gain 3 *l.* 7 *s.* 6 *d. in* 9 months, after the fame Rate.

A PAGE FROM "COCKER'S ARITHMETICK" (1688 EDITION)

A problem in the "Double Rule of Three Direct." The Rule of Three was a semi-mechanical method of solving problems such as are dealt with in this section. The phrase "according to Cocker" derives from this Arithmetic which was very popular for many years.

One of the tractors is out of order. How long should the other three take?

5. What is the cost of eggs per dozen when they are 7s. 1d. per score?

6. James was 4 ft. tall at the age of 12 years. How tall should he be when he is 48 years old?

7. Twelve men were left in a boat, after their ship had sunk, with enough rations for thirty days. If they were joined by another three men, how long would the rations last if the daily allowance was kept the same?

8. A car does a journey of 84 miles on $3\frac{1}{2}$ gallons of petrol. There are still $5\frac{1}{4}$ gallons left in the tank. How much farther can the car go without refilling at the same rate of consumption?

9. A coal-merchant has enough coal to supply forty-two customers for twelve days. How long will it last sixty-three customers, without cutting down the allocation per customer?

10. A cog-wheel of 20 teeth engages one of 15 teeth. The first cog makes 150 revs. per minute. How many revs. per minute does the other cog make?

MAKING COMPARISONS

It is not generally realized that we have two quite different ways of making comparisons. Thus, in comparing the lengths of two objects it may be said that one of them is 3 ft. longer than the other; or it may be that one of them is three times as long as the other. Each statement may be perfectly true; but, generally speaking, the second is more informative. If one of the objects was 100 ft. long the other would be 103 ft. long according to the first method of comparison; according to the second way the other object would be 300 ft. long. In the first way the important words are 'longer than'; in the second way 'as long as.'

All our weights and measures really contain this idea of 'as long as' or 'as heavy as' or 'as much as.' For when we give an object a length, or a weight, we are really comparing it with a unit length such as an inch, or a yard, or a mile; or with a unit weight like a lb. or a ton; and so on.

We call a comparison of the second kind—a 'how many times' comparison—a *ratio*. Thus we say that 6 ft. is $\frac{3}{2}$ times as long as 4 ft. (half as long again), or that the ratio of 6 ft. to 4 ft. is as 3 is to 2, and we write it as:

$$\frac{6 \text{ ft.}}{4 \text{ ft.}} = \frac{3}{2}, \text{ or } 6 \text{ ft.} : 4 \text{ ft.} = 3 : 2.$$

The ratio 3 : 2 is read 'three is to two.'

In finding the ratio of two quantities we just express each in terms of the same units, and then we are in a position to compare them.

Thus, if I have 17s. 6d. and you have 7s. 6d., we can say:

17s. 6d. = 210d.;   7s. 6d. = 90d.

So the ratio of 17s. 6d. to 7s. 6d. = 210 : 90.

But 17s. 6d. also = 7 half-crowns, and 7s. 6d. = 3 half-crowns.

So the ratio may also be given as = 7 : 3.

$$\text{Now } \frac{210}{90} = \frac{7}{3}, \text{ on cancelling down.}$$

This justifies the usual practice for finding the ratio of two quantities by which we bring each to the same units, form a fraction using these quantities as numerator and denominator, and cancel down the fraction to its lowest terms. For example:

What is the ratio of 8 ft. 4 in. to 2 ft. 6 in.?

8 ft. 4 in.   = 100 in.
2 ft. 6 in.   = 30 in.

$$\text{So the ratio } = \frac{100}{30} = \frac{10}{3} \text{ (or 10 : 3).}$$

Note. The words *ratio* and *rate* must not be confused. Both imply division; in the case of ratio, however, the quantities divided must be expressible in the same units; in the case of rate, quantities of different kinds have to be divided. Thus, if a distance of 150 miles is covered in 5 hours this is an average rate of 30 m.p.h.; and if it is done on 6 gallons of petrol this means a consumption of 25 m.p.g. These are rates. We can only give as ratios a comparison between two distances, or two speeds, or two petrol consumptions, and so forth.

Ratios are often given in the form 10 : 1, or—as for gear ratios—3·6 : 1. This is done by reducing the ratio-fraction to an equivalent fraction with denominator 1.

Thus the ratio 10 : 3 which we found above may be expressed as $3\frac{1}{3}$: 1, or 3·3 : 1, to one place of decimals.

In some cases even the 1 is omitted. In the case of an aeroplane, for example, a certain ratio of lengths (the aspect ratio) will be given as, say, 7·7, and not as 7·7 : 1.

EXERCISE 15

1. Give three pairs of numbers whose ratio is 2 : 1.
2. Give three pairs of numbers whose ratio is 3 : 4.
3. Compare, where possible, the following, giving the answers as the ratio of the first to the second in each pair: (a) £1 and 3s. 6d.; (b) quarter of a mile and 100 yd.; (c) 60 miles and 2 gallons; (d) 6 in. and 1 mile; (e) 10 stone and 5 ft. 6 in.
4. An American ton is 2000 lb. What is the ratio of a British ton to an American ton?
5. A nautical mile is 6080 ft. 1 knot is a speed of 1 nautical mile per hour. Compare a speed of 33 knots with a speed of 36 m.p.h.
6. It is said that a window gives the most pleasing effect if the ratio of its height to its width is 7 : 5. (a) How high should such a window be if its width is 3 ft. 4 in.? (b) What width should go with a height of 5 ft. 10 in.? (c) Does a window 8 ft. by 6 ft. satisfy the requirements?
7. The second gear of a certain 3-speed gear is advertised as giving a 25 per cent. reduction from normal speed. What is the ratio of engine speed in second gear to that in normal gear for the same road speed?

8. A six-bushel corn-bin is advertised for £2 10s.; an eight-bushel bin of the same kind costs £2 18s. Is the price of the bins in the same ratio as their capacities—in other words, are they proportional?

PROPORTIONAL DIVISION

This is an extension of the ratio principle. It is best illustrated by an example which is of frequent occurrence.

Mr X in his will directed that his net estate, which amounted to £4500, was to be divided between his three children, John, Harry, and Mary, in the ratio 6 : 5 : 7. How can this be done?

This is the same as giving 6 shares to John, 5 to Harry, and 7 to Mary. In all, there are 6 + 5 + 7, or 18, shares.

$$\text{So John gets } \frac{6}{18} \text{ of } £4500 = £1500.$$

$$\text{And Harry gets } \frac{5}{18} \text{ of } £4500 = £1250.$$

$$\text{And Mary gets } \frac{7}{18} \text{ of } £4500 = £1750$$

$$\text{Check } = £4500$$

Never forget the check. The shares should add up to the total amount divided.

EXERCISE 16

1. Brown, Smith, and Robinson rent a field for £12 a year. Brown puts in 8 cows, Smith 10 cows, and Robinson 12 cows. How should the rent be divided among them?

2. Two men agree to share expenses which amount to £3 15s. in the ratio 7 : 3. How much will their shares be?

3. A sum of five guineas is to be given in three prizes. The first prize is to be double the second prize, and the second prize is to be double the third prize. What will the prizes be?

4. Can you divide £122 5s. 8d. into an equal number of pounds, shillings, and pence?

5. Three men take lodgings together and agree to pay for their food in proportion to the number of meals taken by each. In one week of six days the total bill came to £2 0s. 6d. They agreed that the costs of breakfast and supper (they have dinner and tea out) should be reckoned in the ratio 2 : 3. They were all in for breakfast for the six days; the first man was in for supper each night, the second man missed supper once, and the third man twice. How much should each pay as his share of the meals?

INTEREST

The practice of paying interest is probably as old as money itself, for charging interest for the loan (that is, the use) of money is as reasonable as charging rent for the hire of property.

From the earliest times there is evidence of legal regulation of the rate of interest to be charged. At Rome in 450 B.C. the legal rate seems to have been one-twelfth of the capital per annum. About 80 B.C. the legal rate was 1% per month or 12% per annum. (We have previously noted the tendency of the Romans to work in twelfths.)

At one time the practice of charging interest on loans was denounced by the Church as sinful. But by 1515 the Church realized the futility of its position, and permitted the charging of a moderate interest. In the time of Henry VIII interest was at 10%; a hundred years later it had dropped to 8%. Nowadays the borrower has the courts in which to sue the extortionate lender; and the Government, by issuing securities bearing interest at fixed rates of $2\frac{1}{2}\%$ or 3%, sets a standard of interest which in practice regulates or controls other rates of interest.

The interest according to which most reckonings are made is called *simple interest*. If interest is paid at the rate of 3% per annum then the interest on £100 for one year is £3. It is clear that the principle of direct proportion operates here; the interest paid increases or decreases directly in proportion to the money on which it is paid. Similarly, the interest paid is in direct proportion to the number of years for which it is paid—double the number of years, double the interest, and so on.

The reader will be able to see for himself that the interest on a sum of money (generally called the *principal*) may thus be reckoned from:

$$\text{Interest} = \frac{\text{principal} \times \text{rate} \times \text{times in years}}{100}.$$

Thus the interest on £350 for 4 years at 3%

$$= \frac{£350 \times 3 \times 4}{100} = £42, \text{ on cancelling down.}$$

There are other ways of reckoning interest. For example:

$$5\% \text{ of } £1 = \tfrac{5}{100} \text{ of } £1 = 1s.$$

So a rate of 5% per annum is equivalent to an interest of 1s. in the £ per annum. Similarly, a rate of $2\frac{1}{2}\%$ per annum is the same as 6d. in the £ per annum, or $\frac{1}{2}d.$ in the £ per month. This is how interest is calculated in the Post Office Savings Bank, where the rate is $2\frac{1}{2}\%$ per annum. As a matter of fact, the Post Office Savings Bank calculates interest:

111

(a) only on complete numbers of £; and

(b) only for whole months commencing with the first of a month.

Thus the interest on £28 13s. 6d. from January 5 to April 18 would be reckoned as the interest on £28 for the months of February and March only.

INTRODUCING A FORMULA

We have seen that to calculate the simple interest on a sum of money at a given rate for a given time we multiply the sum of money (the principal) by the rate per cent. per annum and by the time in years, and divide the figure obtained by 100. But it is a very long business to write out all these words every time we wish to calculate simple interest. So to save time, for the simple interest we write the letter I, the principal we call P, the rate per cent. per annum r, the time in years t. We can then write quite shortly:

$$I = \frac{P \times r \times t}{100};$$

or, more shortly still:

$$I = \frac{Prt}{100}.$$

We call this a *formula* for obtaining I, and we say that this formula expresses I in terms of P, r, and t. We shall have a great deal to say about formulæ later. For instance, we shall see how we can change this formula into one which gives us P or r or t in terms of the other letters.

But our immediate purpose here is to show how this use of letters for some numbers enables us to give a general expression which covers all possible numerical cases. It is what we call an arithmetical generalization; it is our first step towards algebra.

APPLIED ARITHMETIC

The working out of the actual simple interest for a particular principal for a given number of years at a given rate per cent. is called 'substituting' in the formula, for we 'substitute' for the letters the particular figures which occur in a given problem.

EXERCISE 17

1. Calculate the simple interest in each of these cases:
   - (a) £350 for 2 years at 4%.
   - (b) £75 for 3 years at 3%.
   - (c) £22 10s. for 4 years at 2½%.

2. Calculate the Post Office Savings Bank interest in the following cases:
   - (a) £25 from March 18 to September 12.
   - (b) £18 7s. 6d. from April 4 to December 2.
   - (c) £12 4s. 7d. from January 12 to December 14.

3. Mr X is charged 6% by his bankers on amounts overdrawn in his account. During one half-year he was overdrawn £30 for four months. What should the charge for this be?

SHARES AND STOCKS

To complete our rapid survey of applied arithmetic let us try to get at the essentials of stocks and shares.

In the days of rapid industrial development in this country the old privately owned company tended to expand itself by 'borrowing' money from outside. This was done by issuing 'stock' or 'shares'—certificates indicating that their holders had, so to speak, lent to the company the amount of money inscribed on them.

Stocks are generally issued in multiples of £100, and nowadays almost always refer to Government issues; shares may have any value from 6d. upward. Stocks and shares can be of different kinds, each kind ranking differently for the distribution of the profits (if any) in the form of 'dividends,' or interest.

In the financial pages of our daily newspapers appear such items as the following:

MISCELLANEOUS

Security	Market Price	Change
Assd. Br. Pc. . .	24/–	– 3d.
Gaumont British .	23/3	+ 3d.
Odeon . . .	41/3	..
Ranks (5s.) . .	20/3	+ 3d.

In the second column appears the market price. It is understood that if no indication to the contrary is given (as in the Ranks 5s. shares above) this is the market price of a £1 share—that is, the day-to-day value of an investment of £1 in the company. If the company is doing well, and paying regular dividends, this may well be over £1; if not, it may be considerably less than £1.

The third column gives the previous day's change in price (if any), + denoting a rise, – a fall, as we have seen when dealing with signed numbers.

Let us now see how the 'yield' (or interest, or dividend) on the money actually paid for a share is calculated.

FINDING THE YIELD OF A STOCK OR SHARE

At the time when these words were written Debenham's 7% 10s. preference shares stood at 17s. What was their yield per cent.?

Actual yield per share
$$= 7\% \text{ of } 10s.$$
$$= \frac{7}{100} \text{ of } 10s. = \frac{7}{10}s.$$

114

$$\text{Yield per cent.} = \frac{\frac{7}{10}}{17} \times 100 = \frac{7}{10} \times \frac{1}{17} \times \overset{10}{100}$$

$$= \frac{70}{17} = 4\frac{2}{17} \text{ (or } 4 \cdot 12\text{)}.$$

So, although they are 7% shares, the yield on money invested is only $4\frac{2}{17}$ per cent.

Of course, there are other expenses in buying and selling stocks and shares, but the above gives a way of comparing the return on money invested.

EXERCISE 18

1. What is the yield per cent. on a 4% stock when it stands at 120?

2. If the first two companies quoted at p. 114 paid a dividend of 7% on their shares what was the yield per cent. to an investor?

3. A man sold out £200 2½% Consols when they stood at 84, and re-invested the money obtained in Ranks' 5s. shares when they were at 20s. If Ranks paid 16% what was the change in his income (nearest penny)? Assume that he paid the selling and buying charges without touching the capital.

4. An issue of cinema shares was made in which the public was invited to subscribe for 4% £1 shares at 21s. per share. What would be the yield on these shares?

The financial columns of newspapers will provide plenty more practice.

CHAPTER VIII

GRAPHS AND HOW TO READ THEM

WHAT IS A GRAPH?

In our discussions on number-form we cannot neglect pictorial methods of recording and comparing numbers. Nowadays we are becoming quite accustomed to seeing in our daily newspapers and on advertisement hoardings diagrams or pictures to illustrate numerical facts—man-power figures, electricity consumption, accident statistics, output diagrams, variation in population, and so on.

In our hospitals records of patients' temperatures are shown on squared paper; and a point of daily pilgrimage at seaside resorts is the barograph in its glass-sided case, where the amateur weather forecaster seeks information, and, if possible, confirmation, for his weather prophecies.

All these pictorial methods illustrate different kinds of *graphs*. The word 'graph' comes from a Greek word meaning 'to write'; and graphs are another way of writing down numbers. More than that, they afford a very convenient way of comparing numbers and of recording variation.

There are many kinds of graphs; for the present we shall limit ourselves to statistical graphs and graphs which can be used for shortening calculations. Besides these there is a vast field of graphs used in connexion with algebra. We shall consider some of them at a later stage.

Statistics may be illustrated graphically in four different ways: (i) by pictographs, (ii) by circle graphs, (iii) by bar graphs, and (iv) by continuous curved-line or jagged-line graphs.

Let us take them in turn and see how to get the facts from them.

PICTOGRAPHS

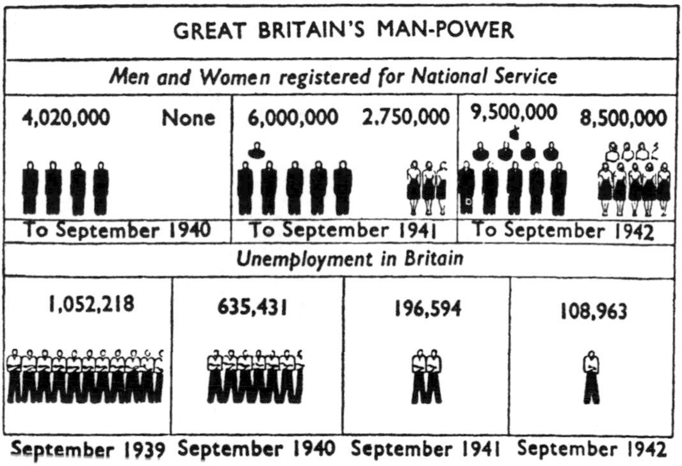

GREAT BRITAIN'S MAN-POWER					
Men and Women registered for National Service					
4,020,000	None	6,000,000	2,750,000	9,500,000	8,500,000
To September 1940		To September 1941		To September 1942	
Unemployment in Britain					
1,052,218		635,431		196,594	108,963
September 1939		September 1940		September 1941	September 1942

Fig. 18

WAR DAMAGE THROUGH ENEMY BOMBINGS BETWEEN SEPTEMBER 1939 AND MARCH 1942

One out of five houses was damaged or destroyed.

Fig. 19

The comparisons of the little drawings tell us quickly what the number facts attached to them represent. For example, in Fig. 20 each man represents 10,000 men.

A very modern development of the pictograph is the *isotype chart,* a specimen of one of which is reproduced at p. 123.

Truly there is nothing new in the world. Thousands of years ago the Arabs and the Egyptians used little

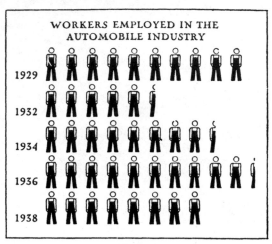

FIG. 20

pictures to represent numbers. We seem to have turned a full cycle back to their methods when we use pictographs!

CIRCLE GRAPHS

FIG. 21

This is a simpler though perhaps not so interesting a way. It depends upon the relative sizes of the parts into which a circle is divided.

Fig. 21 shows one plan for dividing up one's income. Notice how the following questions are answered from the graph.

(i) About what part is allowed for each division?

118

Food, $\frac{1}{4}$; pleasure, $\frac{1}{8}$; savings, $\frac{1}{8}$; clothes, $\frac{1}{8}$; rent, light, and heat, $\frac{3}{8}$.

(ii) How does the amount for pleasure plus savings compare with that for food?

They are equal.

(iii) If the total net wages are £4 per week, about how much does this plan provide for each item?

Using the fractions in (i), we have:

Food, £1; pleasure, 10s.; savings, 10s.; clothes, 10s.; rent, light, and heat, 30s.

(iv) What percentage of the wages is saved?

$12\frac{1}{2}\%$ (turning $\frac{1}{8}$ into a percentage).

BAR GRAPHS

Here the facts are shown and compared by the lengths of straight lines, or bars. These may be upright or flat.

Fig. 22 shows how consumption of oil varied with the condition of a certain engine.

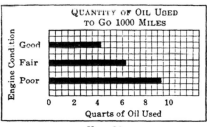

FIG. 22

Look at this bar graph and see how it answers these questions:

(i) About how many quarts of oil are used by the engine in fair condition?

$6\frac{1}{4}$ qt.

(ii) How many more quarts of oil are used by the engine in poor condition than in good condition? How much does this mean when oil is 2s. per quart?

5 qt.; 10s.

(iii) Compare the amount of oil used when the engine

119

is in poor condition with the amount when it is in fair condition.

$$\text{Ratio} = \frac{9\frac{1}{4}}{6\frac{1}{4}} = \frac{37}{25} = \frac{1 \cdot 48}{1}$$

MODIFIED BAR GRAPHS

The expenses of owning and driving a car.

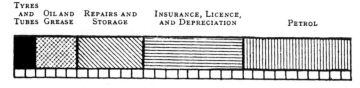

TYRES
AND   OIL AND   REPAIRS AND   INSURANCE, LICENCE,
TUBES   GREASE   STORAGE   AND DEPRECIATION   PETROL

FIG. 23

See how this answers these questions:

(i) Which is the largest item? Petrol.

(ii) What fraction of the total cost is given to repairs and storage? $\frac{6}{31}$.

(iii) What percentage of the total cost is given to petrol? Fraction $= \frac{10}{31}$. Percentage $= 32$ (about).

CONTINUOUS AND JAGGED-LINE GRAPHS

In the top part of Fig. 24 there are eleven thermometers side by side against a background of squared paper, showing how the mercury stood in them each hour from 9 A.M. to 7 P.M. on a certain day. (The pieces of wood on which they are mounted are not shown, and the upper parts of the thermometers are cut short.) Glancing over them, the eye at once takes in how the temperature varied during the course of the day.

If we wished to keep a record of temperatures this would be rather an inconvenient way of doing it. We are chiefly concerned with the actual temperature registered at each hour—that is, only with the *top* of each mercury column.

Now look at the lower part of Fig. 24, where the temperatures are shown in a different way. The thermometers themselves are left out, and two straight lines are

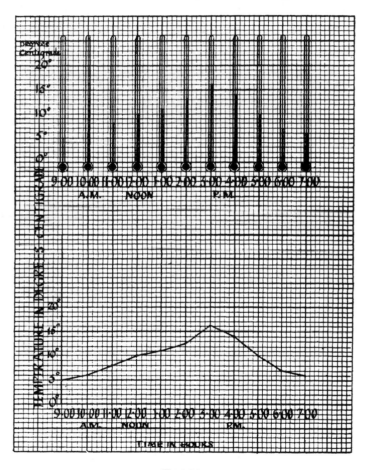

FIG. 24

ruled at right angles to each other. The horizontal line gives the times, and the upright line the temperatures, as at the bottom and side of the thermometer pictures above. The heights of the mercury at the different times may be marked by a dot or small cross. If we join each point to the next one we obtain a picture, or graph, of how the temperature varies, and, of course, the temperature at any time is given by the position of the corresponding point.

In this graph we have joined the points by straight lines, thus making what we have called a jagged-line graph. Probably a continuous curve should have been drawn to pass through them, for temperature does not change by jumps.

One great advantage of a continuous-line curve is that we can read off from it values lying between those already known or given. For example, if we had here drawn a continuous-line curve instead of a jagged-line one we should have been able to read off the temperature at any time between the hours, and the time or times at which any given temperature was registered.

One more thing: the straight lines which we draw at right angles to each other are called the *axes* (plural of axis) of the graph, and the point at which they meet is called the *origin*.

To read this graph (or any other graph) the axes must first be examined so that the scale used may be properly understood. In most graphs the axes are clearly labelled and/or numbered; in many graphs the scales are also shown separately. The vertical axis is generally used for the quantities whose variation is to be shown.

To return to our temperature graph: if the temperature at any particular time is required—say, at 4 P.M.—we first find 4 P.M. on the time axis (here the cross-axis). We then follow the upright line through this point until it meets the graph. From there we follow the cross-line as far as the temperature axis. The reading is 14°.

If the time when the temperature was 11° is required we first find 11° on the temperature axis, and follow the cross-line through this point until it meets the graph.

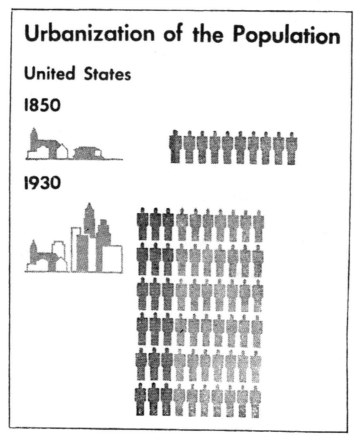

AN ISOTYPE CHART

From *America and Britain*, by L. S. Florence and K. B. Smellie
(Harrap, 1946).

In this case the cross-line meets the graph in two points. Following the upright lines through these points down to the time axis, we have 1 P.M. and 4.48 P.M. (each small square = 12 min.) as the required times.

We note also from this graph that:

(i) The temperature was highest at 3 P.M.

(ii) The temperature increased most sharply between 2 and 3 P.M.

(iii) The lowest temperature recorded was at 9 A.M.

You should be able to make other deductions yourself.

When drawing a graph the following ideas should be kept in mind:

(i) Use as large a scale on each axis as your graph paper allows; but

(ii) Always choose a 'handy' number of small or large squares for units; for instance, three small squares to an inch is not a convenient way of representing lengths given in decimals of an inch. If possible, choose five or ten small squares for this.

(iii) Mark the units clearly on the axes.

(iv) Use the vertical axis for quantities whose variation is to be shown.

(v) The scales on each axis need not be the same.

TIME-SAVING GRAPHS

From the many types of graph which are used to shorten calculations space permits us to select only a few. It is hoped that they will serve as pointers for interested readers.

PERCENTAGE GRAPHS

A merchant wishes to mark his goods so that the selling price shows a profit to him of 30 per cent. on the cost price.

The calculation is not difficult, but good working answers may be obtained at sight by a graphical method. We shall consider cost prices up to 40s.; this will mean selling prices up to $40 \times \frac{130}{100}$, or 52s.

On a sheet of graph paper two axes are drawn at right angles to each other. With a scale of 1 small square = 1s. on each axis, cost prices up to 40s. (i.e., 4 in.) are marked on the cross-axis and selling prices up to 60s. (i.e., 6 in.) on the upright axis.

We have found that a selling price of 52s. corresponds to a cost price of 40s. It follows that a selling price of 26s. corresponds to a cost price of 20s., 13s. to 10s., and so on down as far as 0s. to 0s.

We next mark by small crosses the points depicting cost price 10s., selling price 13s., etc. We denote these points conveniently as (10, 13), (20, 26), (40, 52), . . . (0, 0). Technically these are known as 'number-pairs'; each of them fixes the position of a point by reference to a pair of axes at right angles to each other. Marking the points is generally called 'plotting.'

To 'plot' the point (10, 13) we find 10 (the first number of the number-pair) on the cross-axis; we then move along the upright line through the 10 until we reach a position opposite 13 (the second number of the number-pair) on the upright axis.

If the points (10, 13), (20, 26), (40, 52) are plotted we shall find that they all lie on a straight line, and that the straight line passes through the origin (0, 0). This is not an accident. It is found that when the numbers of the number-pairs are in direct proportion to each other (as is clearly the case here) the points representing them lie on a straight line through the origin.

This introduces us to an important principle in mathematics, and indeed in other sciences. It is that as all the 'plotted' points (and we can easily increase the

125

number of them) lie upon a straight line, then all the possible intermediate points must also lie on the same straight line. This is our justification for 'reading' from

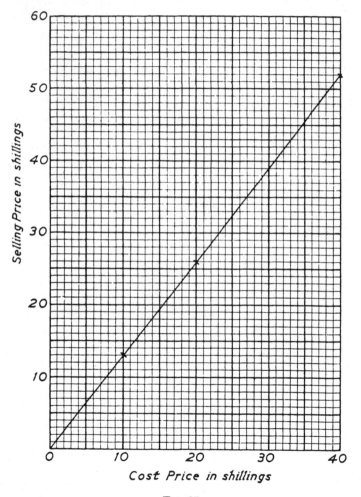

FIG. 25

the graph corresponding values at intermediate points. This is the principle of *interpolation*. It is the basis of much scientific work.

There is also a corresponding process called *extrapolation*. This, in brief, is the reading of corresponding values at points which lie outside the range over which the graph has actually been drawn. In the graph which we are considering this would be quite valid; but, generally speaking, though interpolation may safely be done, extrapolation is likely to be more risky.

Having drawn our graph, we can use it (in the way explained for the temperature graph at pp. 122–124) to solve such problems as the following:

(*a*) What should the selling prices be corresponding to the following cost prices:
    30*s.*, 27*s.*, 34*s.*, 6*s.*, 15*s.*?
Answers: 39*s.*, 35*s.*, 44*s.*, 8*s.*, 19*s.* 6*d.* (approximately)

(*b*) To what cost prices do the following selling prices correspond:
    40*s.*, 21*s.*, 17*s.*, 48*s.*, 32*s.* 6*d.*?
Answers: 31*s.*, 16*s.*, 13*s.*, 37*s.*, 25*s.* (approximately)

EXERCISE 19

1. Make a graph to find selling prices giving a profit of 15 per cent. on goods whose cost price ranges up to £5.
Find (*a*) the S.P. corresponding to C.P.'s of £2, £3 5*s.*, £4 10*s.*, 15*s.*;
    (*b*) the C.P. corresponding to S.P.'s of £2, £3 5*s.*, £4 10*s.*, 15*s.*

2. A tradesman allows a discount of 18 per cent. on marked prices. Make a graph to show actual selling prices for goods whose marked prices range up to £4. From it find:
    (*a*) the selling prices of goods marked at £3, £2 10*s.*, £3 15*s.*, £1 7*s.* 6*d.*;
    (*b*) the marked prices of goods whose actual selling price is £2, £2 10*s.*, £1 15*s.*, 12*s.* 6*d.*

3. A graph which is constructed on a similar principle:

By drawing the straight line joining the origin to the point (8, 25·6) we have a graph which performs multiplications and divisions by 3·2. (8 × 3·2 = 25·6.) From the graph (a) multiply the following numbers by 3·2: 6·5, 4·8, 3·2, 7·9, 4·7; (b) divide the following numbers by 3·2: 20, 10, 15, 23·5, 18·5. (Answers to one place of decimals.)

4. *Conversion* graphs belong to the same class.

8 kilometres = 5 miles (roughly). So if the origin is joined to the point (8, 5) we have a graph from which we can read conversions from kilometres to miles, and *vice versa*. Draw such a graph, and use it to convert

(a) 59, 18, 52 kilometres to miles (nearest mile);
(b) 10, 4·7, 48 miles to kilometres (nearest half-kilometre).

5. A car travels 50 miles in 73 min. By joining the origin to the point (73, 50) find:

(a) its average speed in m.p.h.;
(b) how far it will go in 44 min.;
(c) how long it will take to travel 32 miles.

CURVED-LINE GRAPHS

We have already seen that some number-facts, such as variation in temperature, are best displayed graphically by means of a continuous (that is, a smooth) curved line. Here is another, and a very topical, example. The overall distance required to stop a car on a level road at different speeds is given in the following table. In the compilation of these figures it has been supposed that the driver of the car is alert, that the road surface is good, and that the brakes are sound.

Speed in m.p.h.	10	20	30	40	50	60
Stopping distance in ft.	15	40	75	115	165	225

The reader should first note the scale used on each axis and then check the plotting of the points. The positioning of the points suggests that a smooth curve is the best way of joining the points to form the graph; the principle of interpolation may accordingly be applied.

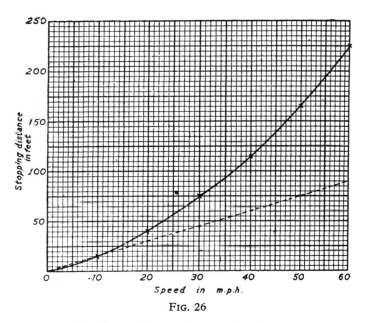

FIG. 26

Some significant factors in road safety may be deduced from this graph. It will be observed that the distance necessary for stopping a car increases sharply as the speed increases. The broken line which has been drawn through the origin and the point (10, 15) shows what the distance would be if it increased in direct proportion to that necessary at a speed of 10 m.p.h. An examination of the upright lengths between the straight line and the curve will teach any road-user a salutary lesson.

I   129

EXERCISE 20

1. From the graph read:
   (a) the distance necessary to stop a car travelling at the following speeds: 26, 58, 34, 15 m.p.h.,
   (b) the probable speed at which a car was travelling if it was pulled up in 50, 125, 200, 100 ft.

2. The broken line in the graph shows what the stopping distance would be at various speeds if it were always directly proportional to the stopping distance at 10 m.p.h. By how much is this stopping distance actually exceeded at 20, 30, 40, 50, 60 m.p.h.?

Here we leave graphs for the present, with just a hint of what is yet to come. We have seen how number-facts may be shown in tabular form, and how they may often be more conveniently pictured graphically. We shall learn presently that there are two more aspects of number-facts which we shall need to study. A relation may often be shown by what is called a formula in algebra— and this will bring in what is described as functional relationship; and, further, many formulæ can be displayed graphically.

PART II

LINES—LINES—LINES

CHAPTER IX

FUNDAMENTAL IDEAS IN GEOMETRY

THE ORIGINS OF GEOMETRY

WE have already mentioned how man, thousands of years ago, in his search after beauty developed the most intricate geometrical designs. But there was also a still more practical side to his use of geometry. In ancient Egypt the people outside the cities lived in small villages which were situated on mounds and shaded by palms. Their farmlands were on the level lowland plain, and when the river Nile rose in flood it deposited a layer of mud over the country which wiped out the boundary marks between the fields. Incidentally, the people still live there, and the Nile continues to perform its useful function.

The Greek historian Herodotus (484–425 B.C.) tells us in the second book of his *History* that the land of Egypt was divided up into square plots of equal size and let to the holders for a yearly rent. He goes on:

> But every one from whose part the river tore anything away had to go to him [the King] to notify what had happened; he then sent overseers, who had to measure out by how much the land had become smaller, in order that the owner might pay on what was left, in proportion to the entire tax imposed. . . . In this way, it appears to me, geometry originated, which passed thence to Greece.

131

Indeed, it is from this that we have the name 'geometry,' from the Greek words meaning 'land measuring.'

There was another important way in which the Egyptians, and also the Babylonians, used what we to-day call practical geometry. Their temples had to be placed so as to face certain definite directions. By determining the direction of the rising and the setting sun at the equinox, and drawing a line at right angles to this, they had a line which pointed due north and south. In our own Stonehenge a certain stone marks the day of the summer solstice, when the sun rises farthest north along the eastern horizon.

The Greeks, always more interested in theory than in practice, developed geometry into a powerful intellectual instrument. Later on we shall consider their contribution to theoretical geometry. But first we must be sure that we understand the language of geometry—in other words, the meaning of the common geometrical terms.

BASIC IDEAS IN GEOMETRY
(This section may be omitted at a first reading)

We start with the idea of a point. The old definition says, "A point has position but no magnitude." It is a perfectly good definition, but rather top-heavy for such a simple idea. For a point is just a mark used to fix a particular position.

We often make appointments in such general terms as, "I'll meet you under the clock at Victoria Station." This is another way of expressing the same idea, but it has not the precision of the mathematical definition of a point. This suggests a place—but it is not dependent upon the size of the place of meeting. It just marks a position.

A line and a surface may be defined in a like manner. But we shall better clarify our ideas if we start from the

132

ONE ASPECT OF THE MATHEMATICAL
IDEA OF ABSTRACTION
(See p. 135)

known, or tangible, and work backward to the unknown, or intangible.

We all have some idea of what a solid is. Now a solid has an inside, so to speak, and an outside; between the two comes a 'surface.' So a surface separates any portion of space into two parts; it does not belong to either part. Continuing, we say that if a surface is divided in any way into two parts the separating boundary is a line. The line does not properly belong to either part. Lastly, we think of a point as separating two parts of a line.

These ideas fit in well with the notions that two lines which meet or intersect do so in a point, and that two surfaces which meet or cross each other do so in a line.

There is one kind of surface which is of especial importance to us: a plane, or flat, surface. We define a plane (surface) as one in which, if any two points are taken, the straight line joining those points lies wholly in the surface. How does a workman lay a flat cement surface? He finishes off his surface by drawing a straight-edge, generally the edge of a board, across it while it is still soft.

Of course, we have to picture a line as a continuous mark, for we have to be able to draw a line in some way. (The idea of a boundary or separation may perhaps be better seen in a crease when paper is folded.) And we picture a surface as something tangible—the top of a table or the outside of a ball, for instance. But we also speak of the surface of a lake or even of a cloud.

The old-fashioned definition of a *straight* line was "the shortest distance between two points." A mariner or an airman might have his doubts about this. The definition generally accepted now is "a line that keeps the same direction from point to point throughout its whole length." This leads to the definition of a curved line as "a line that changes its direction from point to point."

At this stage we can with profit indicate one aspect of the mathematical idea of abstraction. Imagine a person standing on high ground with an ordnance survey map of the district set out before him. He sees all round him roads, rivers, perhaps a railway, and so on. The roads may be winding or straight—they will be marked on the map as winding or straight lines. Some of the roads may be leafy glades, pleasant to the eye and restful to the mind; others may be completely devoid of interest. The lines drawn to represent the roads on the map are not expected to convey this. The map-maker 'abstracts' what is purely practical from the picture of the road and shows it merely by a line. The length and direction of the road from Barchester to Framley does not depend upon whether there is good blackberry picking in the hedgerows or whether the larches on the right-hand side really are delightful in the spring. The map strips the road of all but its bare mathematical essentials; if we know the road the imagination can do the rest.

This idea of abstraction lies at the root of most mathematical practice. Generally speaking, the better an abstractor a person is, the more likely is he to make a competent mathematician. The person who has continually to be limited to the concrete is, from the mathematical point of view, an inferior type!

ANGLES

You will probably have learned that when two straight lines meet they form an angle. The more modern way of regarding an angle is that an angle is formed when a straight line turns about a point in its own length. The point is called the *vertex* of the angle; the first and last positions of the straight line are called the *arms* of the angle.

We name a straight line by writing down next to each

other the letters denoting two points on the straight line. We use three letters for naming an angle; one is the point where the two straight lines meet, and the other two are points one on each of the arms of the angle. The angle shown in Fig. 27 is the angle AOB, generally written as $\angle AOB$, or $A\hat{O}B$. $A\hat{O}B$ is the same as $B\hat{O}A$. The size of an angle does not depend upon the length of its arms.

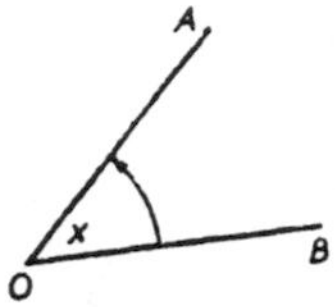

FIG. 27

Where there is no possibility of confusion with other angles at the same point a single capital letter or small letter may be used to denote an angle. Thus $A\hat{O}B$ may be referred to as $\hat{O}$, or even O where there can be no confusing it with the point O; and it may also be referred to by a small letter such as x, the x being marked between the arms of the angle near the vertex.

RIGHT ANGLES: DEGREES

Take a piece of paper; fold it, and crease the fold. The crease forms a straight line—say, AB (Fig. 28). Now fold the paper so that A falls on B, and crease the fold. If the paper is opened up a new crease —say, CD—appears. We say that CD is at right angles to AB, or that CD is *perpendicular* to AB. If AB and CD cut at O, $C\hat{O}B$, $C\hat{O}A$, $B\hat{O}D$, $A\hat{O}D$, are all equal, and each is a right angle.

FIG. 28

A right angle is therefore a quarter of a complete turn round.

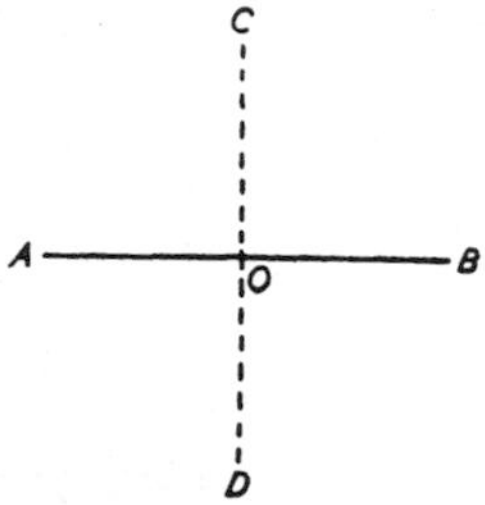

136

The Babylonians noticed that the sun made a complete turn round the earth (as they thought), and they reckoned that this took place every 360 days. So they divided a complete turn about into 360 parts, or degrees—written 360°—and angles have been measured in degrees ever since. The measurement of angles is done by means of a protractor.

As a right angle is a quarter of a complete turn round, a right angle is an angle of 90°. Angles less than a right angle are called *acute* angles; the old name for acute angles was 'sharp' angles. Angles greater than a right angle, but less than two right angles, are called *obtuse* angles. An angle of two right angles (a straight line) is called a *straight* angle.

Two very important directions are the *horizontal* and the *vertical* directions. The vertical direction is the direction in which a plumb-line hangs; the horizontal direction is at right angles to the vertical direction; it is tested by means of a spirit-level.

PARALLEL STRAIGHT LINES

It is evident that straight lines can be drawn on a flat surface which will not meet however far they are produced. The ruled lines on an exercise book are an example. Such straight lines are called *parallel straight lines*. There is no angle between them; they always keep the same distance apart. Note that the definition only applies to straight lines in the same plane. Innumerable instances can be given of straight lines which cannot possibly meet however far they are produced, and yet are not parallel. If you are reading this while sitting in a room you will see at once that the line in which the ceiling meets the wall to your right or left cannot meet the line in which the wall in front of or behind you meets the floor. Yet they are not parallel; they lie in different planes.

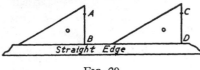

FIG. 29

We generally draw parallels by using set-squares. Fig. 29 illustrates one method of using set-squares. *AB* and *CD* are both perpendicular to the straight edge. There is therefore no angle between them; so they are parallel to each other.

TRIANGLES

Two points *A*, *B*, are joined by a straight line *AB*. Three points which do not lie in a straight line can be joined by three separate straight lines. The joins are said to form a triangle. Strictly speaking, we should call it a trilateral. A triangle is named by its three angular points, or vertices. Thus the triangle formed by joining the points *A*, *B*, *C* is the triangle *ABC*. It is written $\triangle ABC$.

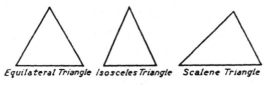

Equilateral Triangle Isosceles Triangle Scalene Triangle

FIG. 30

Triangles with all their sides equal are called *equilateral* triangles. The derivation of the adjective is obvious. Triangles which have two of their sides equal to each other are called *isosceles* triangles (from the Greek *isos*, equal; *skelos*, leg). It can be shown that if two sides of a triangle are equal then the angles opposite them are also equal. Triangles with three unequal sides are called *scalene* triangles (from the Greek *skalenos*, crooked).

138

A PROPERTY COMMON TO ALL TRIANGLES

Draw a triangle on stiff paper or cardboard. Cut off the corners along the straight and wavy lines as in Fig. 31. Fit them together as in Fig. 32. The free arms of the angles A and C will then be found to lie in a straight line, and so $\hat{A} + \hat{B} + \hat{C} = 180°$, or two right angles.

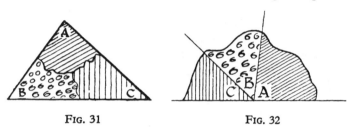

FIG. 31   FIG. 32

Or we may proceed as follows: in Fig. 33 L and M are the mid-points of the sides on which they lie. LX and MY are drawn perpendicular to the bottom side (the *base* of the triangle, as it is called). Cut out the triangle, and fold the corners over the dotted lines LM, LX, MY as in Fig. 34. We see as before that $\hat{A} + \hat{B} + \hat{C} = 180°$, or two right angles.

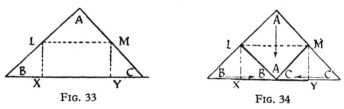

FIG. 33   FIG. 34

We have therefore checked in two ways that *the sum of the angles of any triangle is* 180°, *or two right angles.* This can be proved rigorously.

Note that we have established this important property for *any* triangle of any shape whatsoever. We have therefore found another instance of the generality of mathematics. What we have shown for the particular triangles cut out holds for any triangle and for all triangles.

We can use this *angle-sum property*, as it is generally called, in many ways. From it we can find the third angle of a triangle if two of the angles are known. For example, if two of the angles of a triangle are known to measure 60° and 42° the third angle is 180° − 60° − 42°, or 180° − 102°—that is, 78°. Or again, if the angle between the equal sides of an isosceles triangle is 50° the sum of the other two angles is 130°, and so each of them is 65°. In the same way, if two of the angles of a triangle are each 45° the third angle must be a right angle. Lastly, each angle of an equilateral triangle is 60°.

QUADRILATERALS

Geometrical figures which are formed by joining by straight lines four points in a plane (no three of which are in the same straight line) are called *quadrilaterals*.

Our interest in quadrilaterals lies chiefly in those which have special properties. First, there are rectangles and squares, possibly the most common of all geometrical figures. A *rectangle* is a quadrilateral with all its angles right angles. It will be seen at a glance that the opposite sides of a rectangle are equal; if, in addition, *all* the sides are equal, the rectangle is a *square*. Rectangles and squares are largely used by craftsmen in all trades. The reason is clear. Unless one wants to make a 'crazy' paving, the object which has all its angles right angles is the easiest to fit into any pattern. Irregularly cut stone or brickwork is much more expensive to put up than ordinary rectangular-faced stone or brick, because so much time has to be spent in fitting together the awkward edges.

A valuable property of all rectangles is that the *diagonals*—that is, the straight lines joining opposite corners—are equal. A carpenter who wishes to test whether a window-frame is 'squared' measures the distances between the opposite corners. If they are equal (and, of course, if the opposite sides of the frame are also equal) he knows that all the angles are right angles.

A *parallelogram* is a quadrilateral whose opposite sides are parallel. Two sets of straight railway lines which do not cross at right angles form a parallelogram between their intersection points. A useful property of a parallelogram is that its opposite sides are equal. Parallelograms are not met in practice as much as rectangles, but the extending arm shown in Fig. 35 makes use of a parallelogram to ensure that the tray is always horizontal, since its

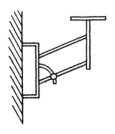

FIG. 35

support is always parallel to the wall in all positions. A parallelogram whose sides are all equal is called a *rhombus*.

A *trapezium*

FIG. 36

There is one more four-sided figure which must be mentioned. It is a *trapezium*. A trapezium has one pair of parallel sides. It is often to be seen in girder work, and a surveyor frequently makes use of trapezia (plural of trapezium) in his field-area measurements.

PLANE FIGURES WITH MORE THAN FOUR SIDES

Names are also given to certain geometrical figures which have more than four straight-line sides. They are derived from the Greek words for the numbers of sides.

141

Thus a *pentagon* is a five-sided figure (*pente*, five, and *gōnia*, an angle), a *hexagon* has six sides, a *heptagon* seven sides, an *octagon* eight sides, and so on.

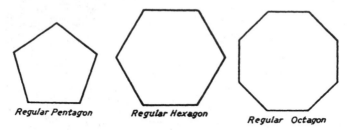

Regular Pentagon   Regular Hexagon   Regular Octagon

FIG. 37

Many-sided figures are most commonly used when all their sides and angles are equal. They are then called *regular polygons* (*poly*, a prefix from the Greek word for 'many'). Many delightful designs—for example, in architecture, carpets, and linoleum—are based on the regular polygons. We shall have occasion to refer to them later.

PLANE FIGURES WHOSE SIDES ARE NOT ALL STRAIGHT LINES

(*a*) *The Circle.* First in this group is the circle. Strictly speaking, a circle is the surface contained within a boundary line which is such that every point in the boundary is at the same distance from a fixed point, called the *centre* of the circle. The boundary is called the *circumference* of the circle. In ordinary speech we tend to say "draw a circle" when what we really mean is "draw the circumference of a circle." A circle is drawn by using compasses, though a piece of thread, string, or rope fastened to a fixed point at one end and having a pencil or stake fastened to the other end does equally well. Any straight line drawn from the centre

142

to the circumference is called a *radius* (plural, *radii*), and a straight line through the centre with each end at the circumference is called a *diameter*. Clearly all radii of a circle are equal, and so are all the diameters.

Draw a circle and a straight line. Imagine the straight line to be moved through a series of parallel positions as in Fig. 38. Then either it will not meet the circle at all (*a*), or it will cut it in two points (*b*), or there are two positions (*c*) in which it meets the circle at apparently only one point. We use the word 'apparently' because what has really happened is that the two points seen in position (*b*) have moved up into

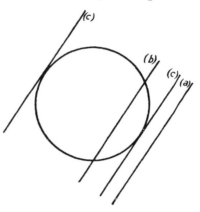

Fig. 38

coincidence in the positions (*c*). In these two positions (*c*) the straight line is said to touch the circle, or to be a *tangent* to the circle. We therefore define a tangent to a circle as a straight line which cuts it in two coincident points. The tangent to any other curve may be defined in the same way.

(*b*) *The Ellipse.* Fasten the ends of a piece of thread (or string or rope) to two points *A* and *B*. Insert a pencil point or a stick (*P*) in the slack of the thread. Keeping the thread taut, you can easily move the pencil or stick in such a way as to trace out a curve. The curve traced out will be something like the one shown in Fig. 39; it is called an *ellipse*. The two fixed points *A* and *B* are called the *foci* (plural of *focus*) of the ellipse.

143

The greatest width of the ellipse (*CD*) is equal to the length of the thread. The depth of the ellipse depends upon the distance *AB*. *CD* is called the *major axis* of the ellipse. It is clear from the manner of construction that the major axis divides the ellipse into two equal halves. Gardeners will find this a useful method of marking out an 'oval' bed.

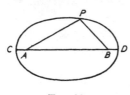

FIG. 39

It is interesting to note that the above method of drawing an ellipse was invented by Clerk Maxwell when a boy. Professor J. D. Forbes was so impressed that he constructed a scientific paper on the idea, and read it before the Royal Society of Edinburgh. Maxwell, not then fifteen, was taken to hear the paper read.

When a circle is viewed from any point not directly above its centre it will look like an ellipse. We make use of this when we have to make a sketch of a solid body which includes circles in its design. The earth (in fact, all the planets) moves in an ellipse about the sun, the sun being at one of the foci. In books on geography and astronomy it is explained how the variation in the length of the seasons is governed by this fact.

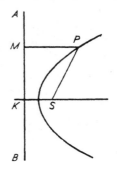

FIG. 40

(*c*) *The Parabola*. Another familiar curve is the *parabola*—the path traced out by a cricket ball when it is not thrown vertically upward. A parabola can be traced out in the following way. Draw a straight line —say, *AB*—and mark a point *S* not on *AB*. Then, if a series of points such as *P* is marked so that *PS* is equal to the perpendicular distance of *P* from *AB*, *P*

144

traces out a parabola (Fig. 40). The point S is called the *focus* of the parabola. If SK is drawn perpendicular to AB, SK is called the *axis* of the parabola. The axis divides the parabola into two equal halves.

Suppose PT is a straight line touching the parabola at P. Join SP, and draw PM perpendicular to AB, producing PM backward to N (Fig. 41). Then it is known that PN and SP make equal angles with the tangent at P. Now wherever P is taken PN will be parallel to the axis of the parabola, since PN and the axis are both perpendicular to AB. If, then, a silvered reflector is made in the shape of a parabola, and a source of light is placed at S, the rays, which are reflected in such a way that they make equal angles with the tangent at the point where they strike the reflector, will be reflected as a beam parallel to the axis.

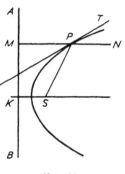

Fig. 41

Of course, it is impossible to get a point source of light, and so this argument may be regarded as theoretical, as anyone who has noticed the diverging and converging rays of light in the beam of a headlamp will at once say. But the principle is important, and use is, in fact, made of it.

The reverse of this is used in sound detectors. If a parabola-shaped sound reflector is hinged in such a way that its axis can be pointed to a source of sound the sound waves which are picked up by the reflector will be directed to S, where they can be picked up by a microphone and amplified.

(*d*) *General Considerations.* The reader may have noticed a certain similarity of idea in the construction of circles, ellipses, and parabolas. A circle is enclosed by a boundary traced out by a point which keeps at the same distance from a certain fixed point. The boundary of an ellipse is such that the sum of the distances of any point on it from two fixed points is always the same. A parabola is traced out by a point whose distance from a certain fixed point is always equal to its distance from a certain straight line. Do you see the connexion—distance from one point (circle); distances from two points (ellipse); distance from a point and a straight line (parabola)? Would you like to complete the series by considering how a point moves if it keeps at the same distance from (i) two fixed points; (ii) two fixed lines? In the second case consider separately two parallel and two non-parallel straight lines. And then, to complete the picture, consider these cases (and cases corresponding to one point, two points, and a point and a line) when you are not restricted to one plane—in other words, when the point moves in space.

A WAY OF CHECKING YOUR KNOWLEDGE OF GEOMETRICAL TERMS

There are several instructive and interesting ways in which this can be done. If you work in a factory look round you—at the machinery, at the roof girders, at the building itself—and see how many of the geometrical forms you can discover. Even while you are waiting for a football match to commence a peep at the girders bearing the roof of the stand will serve the same purpose. Of course, the same sort of thing can be done second-hand, as it were, by studying the pictures in newspapers and elsewhere.

Or again, buildings of architectural value are to be found everywhere. Examine them with a geometrician's eye—the shape of the walls, the roof timbers, the buttresses, the tower, if there is one—there is scarcely a geometrical form which is not to be found in an old building such as a cathedral or church. Modern buildings, too, tend to be severely geometrical—a limitation perhaps imposed by steel and concrete.

GEOMETRY IN ARCHITECTURE
(*For those especially interested*)

Some brief notes are given here of the main features of the different periods of architecture, especially in church buildings. They are based on a geometrical reading of their styles.

NORMAN ARCHWAY

EARLY ENGLISH WINDOW

(*a*) *Norman* (about 1060–1160). Arches and windows with semicircular tops; massive pillars (or piers) of circular section linked by semicircular arches; 'capitals' at the tops of the piers cut out of solid blocks of stone; ornamental work of a stiff geometrical type, chiefly of a zigzag, or chevron, pattern.

(*b*) *Early English* (about 1200–1250). The pointed arch (often based on the equilateral triangle); slender piers, frequently on a hexagonal or octagonal base; windows so acutely pointed that they are often called lancets.

147

(c) *Decorated Gothic* (about 1250–1300). A development of the Early English style; windows with geometrical tracery often founded on equilateral triangles and regular polygons inscribed in circles—that is, with their angular points on circles (*trefoil, quatrefoil, cinquefoil*—three-, four-, and five-pointed).

PERPENDICULAR WINDOW

DECORATED WINDOW

(d) *Perpendicular Gothic* (about 1350–1550). The eye is caught by vertical and horizontal (and so perpendicular) lines everywhere, particularly in the window tracery and the buttresses; flattened arches.

From the time of Sir Christopher Wren (about 1700) there was a revival of the classical forms based on the circle, the triangle, and the rectangle—St Paul's Cathedral is a case in point. More recently a mixture of decorated and perpendicular Gothic has become popular, as in Liverpool Cathedral. Modern domestic architecture—as exemplified, for instance, by blocks of flats, civic buildings, and railway-station façades—makes skilful use of the curve and the straight line.

148

CHAPTER X

WHERE GEOMETRY JOINS ARITHMETIC

GENERAL CONSIDERATIONS

THE *length* of any line is the number of units of length contained in that line. These units may be inches or feet, yards or miles, centimetres, metres, kilometres, or any other measures in use in different countries. The measuring may be done by a rule, a tape-measure, a surveyor's chain; or the length may be calculated by the methods of the surveyor.

Before our measures were fixed by law there were different measures in different parts of the country. For people who counted with their fingers it was natural that they should measure with their fingers and hands and arms and feet. To take only two of our units—a *foot* was originally the length of a man's foot; a *hand* was the breadth of a man's hand. (The height of horses is still measured in hands.) Another interesting unit comes from farming: once upon a time farmers would say that a distance was as long as the furrow made by a plough, or a furrow-long, hence a furlong.

An *area* is the amount of surface included within a closed boundary, and so we have to have different units— units of surface this time. These are usually formed from the squares

FIG. 42

whose sides are the unit length measures. Thus we talk of a square inch—that is, a square whose side is one inch—or a square mile, or a square kilometre, and so on.

We must be careful not to confuse measures of area, such as are contained in expressions like 'a two-inch square' and two square inches. Perhaps these diagrams (which have been reduced in size to fit the page) will help you.

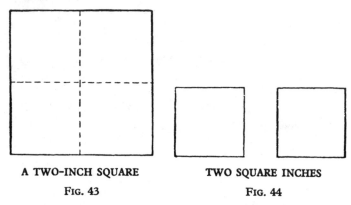

A TWO-INCH SQUARE   TWO SQUARE INCHES

FIG. 43   FIG. 44

There are other units of area which have no direct connexion (so far as their names go) with the units of length, such as acres in this country and the are and hectare on the Continent. By the way, an acre was originally the amount of land that a yoke—that is, a pair —of oxen could plough in a day.

But the point is that measurement of area is expressed in terms of units of area, and the calculation of an area is the calculation of the number of unit squares contained in that area. That is why it often depends upon the multiplication of two numbers. Multiplication may be a quick way of counting the number of unit squares contained in an area.

To help you four diagrams are given illustrating how we obtain the rules for calculating the area of the simplest surface-shapes—the rectangle or oblong (and this includes the square, which is a rectangle with equal sides), the triangle, the parallelogram, and the trapezium. Of course, the area of a surface of any shape can be found by drawing it on squared paper and counting the number of unit squares contained within its boundary; and there are other ways, some simple, some involving more difficult mathematics.

AREAS BY PICTURES

(a) *Rectangle* (*Oblong*)

   Length, 3 in.
   Breadth, 2 in.
   Each square within the broken lines is 1 sq. in. in area.
   Number of squares = 3 × 2 = 6 (or by counting squares).
   Therefore area = 6 sq. in.

If the length of a rectangle is *a* in. (or ft. or metres, etc.) and the breadth is *b* in. (or ft. or metres, etc.), then the area is *a* × *b* sq. in. (or sq. ft. or sq. metres, etc.).

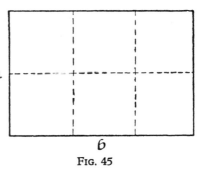

FIG. 45

Note. The multiplication of *a* by *b* (they must be measured in the same units) saves us the trouble of counting the number of unit squares.

From the above the general rule follows that the area of a rectangle is length × breadth.

(b) *Triangle.* Draw a rectangle on the same base and with the same height as the triangle, as in the diagram (Fig. 46).

151

Cut out I and III, and show that after turning they fit exactly on II and IV.

Then the area of the triangle is half that of the rectangle. Since the area of the rectangle is its length multiplied by its breadth, and the length of the rect-angle is the base of the triangle, and the

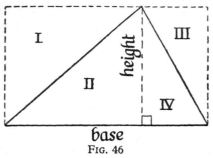

base

FIG. 46

breadth of the rectangle is the height of the triangle, we obtain the following:

The area of a triangle is $\frac{1}{2}$ base × height.

(c) *Parallelogram.* A four-sided figure with both pairs of opposite sides parallel. (They are also equal.)

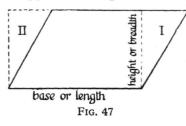

base or length

FIG. 47

Draw the rectangle (sides are the broken lines where they do not lie on the sides of the parallelogram) on the same base and with the same height as the parallelogram.

Cut out I and show that it exactly fits over II.

Then the area of the parallelogram

$$= \text{area of the rectangle}$$
$$= \text{length} \times \text{breadth}$$
$$\text{or base} \times \text{height, so:}$$

The area of a parallelogram is equal to its length × breadth.

Note. Breadth here means height. Units as for (a) above.

152

(d) *Trapezium.* A four-sided figure with one pair of parallel sides.

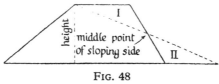

FIG. 48

The construction is shown in the figure.

Cut out I and show that after turning it exactly fits over II.

Area of the trapezium = area of the resulting triangle.
Area of triangle = $\frac{1}{2}$ base × height.
Base of triangle = sum of parallel sides of trapezium;
and height of triangle = height of trapezium
= distance between the parallel sides.

Therefore, the area of a trapezium is equal to half the sum of the parallel sides × the distance between them.

The diagram (Fig. 49) shows the end wall of a factory. It is 200 ft. wide, and the heights to the lowest and highest points of the roof are 118 ft. 6 in. and 151 ft. 9 in. What is the area of the wall in square feet?

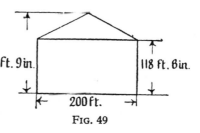

FIG. 49

The wall consists of:

(a) a rectangle with sides 200 ft., 118 ft. 6 in.;

(b) a triangle with base 200 ft., height 151 ft. 9 in. − 118 ft. 6 in., or 33 ft. 3 in.

153

(a) Area of rectangle $= 200 \times 118\frac{1}{2}$ sq. ft.
                     $= 23,700$ sq. ft.

(b) Area of triangle $= \frac{1}{2} \times 200 \times 33\frac{1}{4}$ sq. ft.
                     $= 3325$ sq. ft.

   Total area $= 27,025$ sq. ft.

Fig. 50 is a rough sketch of a park. The lower part is a parallelogram; the upper part is a trapezium. What is its area in acres (rounded off to the nearest acre)?

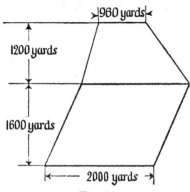

FIG. 50

Area of parallelogram $= 2000 \times 1600$ sq. yd.
                          $= 3,200,000$ sq. yd.

Sum of parallel sides
   of trapezium $= 2000 + 960$ yd.
                         $= 2960$ yd.

So, area of trapezium $= \frac{1}{2} \times 2960 \times 1200$ sq. yd.
                          $= 1480 \times 1200$ sq. yd.
                          $= 1,776,000$ sq. yd.

And area of park $= 4,976,000$ sq. yd. on adding.

Now there are 4840 sq. yd. in an acre.

So area of park $= \dfrac{4,976,000}{4840}$ acres.
                          $= 1028$ acres.

THE CIRCLE

Strangely enough, though the circle is such a simple figure, finding its area is not so easy.

Draw several circles of different sizes. (As a matter of fact, quarter-circles will do.) Count the number of squares contained by the boundary line (or circumference, as it is called). If quarter-circles are used the number will have to be multiplied by four, of course. Half and over-half squares are counted as whole ones, and under-half squares are left out. On the average you will not be far out, and the counting will give the approximate area of the circle. If inch squares subdivided into tenths are used the area will be in square inches to two decimal places.

Now measure the radius of the circle in inches to one decimal place. (The best plan is to make each circle have a radius which is an exact number of inches.)

Divide the area of each circle (in square inches) by the square of its radius, which will also be square inches. You should find that the average

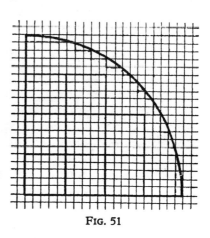

Fig. 51

of these ratios comes out to about 3·14 (roughly $\frac{22}{7}$).

Here is an example. A quarter-circle of radius 2 in. has been drawn. (Fig. 51 is slightly less than full size.)

Number of whole large
         squares = 8, giving   200 small squares.
Number of whole small squares   = 94 small squares.
Half-and-over small squares   = 20 small squares.

                    Total = 314 small squares.

Since each small square $= \cdot 1 \times \cdot 1$ sq. in.
                              $= \cdot 01$ sq. in.
Area of circle $= 4 \times 314 \times \cdot 01$ sq. in.
                              $= 12 \cdot 56$ sq. in.
Now radius^2 $= 2 \times 2$ sq. in.
                              $= 4$ sq. in.
And $\dfrac{\text{area}}{\text{radius}^2}$ $= \dfrac{12 \cdot 56}{4}$
                              $= 3 \cdot 14.$

You should do this for at least two more circles to convince yourself that whatever the radius, the expression $\dfrac{\text{area}}{\text{radius}^2}$ always comes to the same number—that is, about $3 \cdot 14$, or $\frac{22}{7}$.

If you measure the circumferences of your circles, using a piece of cotton, you will also find that the ratio of this length to twice the length of the radius comes to the same number (approximately $3 \cdot 14$).

Or, stand a wheel on a board and mark the wheel and board where they 'touch.' Roll the wheel along a straight line until the mark 'touches' the board again. The distance between the first mark on the board and this point will give the circumference of the wheel. Interesting variations of this would be to use a wheelbarrow or a bicycle-wheel.

This is not an accident; it is a property common to all circles. We cannot find the exact numerical value of this ratio which we found to be roughly $3 \cdot 14$—men have

156

attempted to solve the problem of 'squaring the circle' throughout the ages, but it is given the name π (pronounced 'pi'—the Greek small letter *p*). We say, if the radius of a circle is called *r*, that:

> *The length of the boundary line (the circumference) of a circle whose radius is r* $= 2\pi \times r$.
> *The area of a circle whose radius is r* $= \pi \times r^2$.

Though π cannot be found exactly, it can be calculated to as many places of decimals as we please. Shanks worked it out in 1853 correct to seven hundred and seven decimal places. As a matter of fact, ten places of decimals are enough to give us the circumference of the earth at the equator to within a fraction of an inch, and π = 3·1416 is quite sufficient for really accurate work.

Though mathematicians throughout the ages from the ancient Egyptians, Archimedes and Euclid among them, worked at the value of π, it was left to a man from Anglesey, William Jones by name, to suggest the symbol π for it. This he did in 1706, and in less than fifty years that symbol was adopted everywhere.

The length of a propeller blade from the tip to the centre of the driving-shaft is 5 ft.

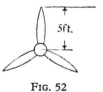

FIG. 52

(*a*) How far does the tip of the blade move during 1000 revolutions of the propeller?

(*b*) What is the area swept out by the propeller? Take π = 3·14.

(*a*) In one revolution the tip of the blade moves along the circumference of a circle of radius 5 ft.

It therefore moves through

$$2\pi \times 5 \text{ ft.} \qquad = 10 \times 3\cdot14 \text{ ft.}$$
$$= 31\cdot4 \text{ ft.}$$

So in 1000 revs. distance moved

    through   = 31,400 ft.

                                    = 6 miles (approx.)

(b) Area swept out $= \pi \times \text{radius}^2$

                       = 3·14 × 25 sq. ft.

                       = 78·5 sq. ft.

Additional Note on Circular Measure. We saw at p. 137 that the size of an angle is measured in degrees. In more advanced mathematics another unit of angle measurement is used.

FIG. 53

If, starting from the end *A* of a radius of a circle centre *O*, a length *AP* is measured along the circumference of the circle equal to the radius (Fig. 53), and *OP* is joined, then the angle *POA* is what is known in circular measure as a *radian.* A radian is a little more than 57°.

Since the length of the circumference of a circle of radius *r* is $2\pi r$, it follows that the arc *AP* is contained 2π times in the circumference of the circle.

So 2π radians = a complete turn = 360°.

Thus a straight angle = π radians;

    a right angle   = $\dfrac{\pi}{2}$ radians;

    an angle of 60° = $\dfrac{\pi}{3}$ radians; and so on.

VOLUMES

We have seen that lengths and areas have their own units. Similarly, volumes have their own units. They are usually formed from the cubes whose sides are the unit length measures—that is, a cube whose sides (or edges) are one inch or one foot or one yard, giving us a cubic inch or a cubic foot or a cubic yard; or, in the metric system, a cubic centimetre or a cubic metre.

There are other units of volume, such as the pint and the gallon, which have no direct connexion with the units of length. But the point is that the measurement of volume is expressed in terms of units of volume, and the calculation of volume is the calculation of the number of units of volume contained in that volume. The diagram shows (if the edges are 5, 5, 3 inches long) that the

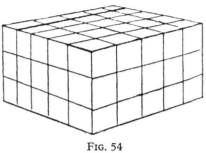

FIG. 54

volume of this rectangular box is 5 × 5 × 3, or 75, cubic inches; the multiplication is a quick way of counting the number of unit cubes contained in the volume.

In general, the volume of a rectangular solid—for example, a box or a room—is $l \times b \times h$, where l, b, h are its length, breadth, and height respectively, all expressed in the same units. Or, if A is the area of the base of the solid and h is its height the volume is $A \times h$. Of course, if h is in feet A must be in square feet, and so on.

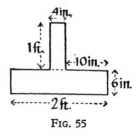

FIG. 55

The diagram illustrates the cross-section of a girder whose length is 36 ft. Find (a) its volume and (b) its weight if it is made of a metal which weighs 200 lb. per cu. ft.

The cross-section consists of two rectangles.

159

Keep all the working in ft., sq. ft., and cu. ft.

(a) Area of cross-section $= 1 \times \frac{1}{3} + 2 \times \frac{1}{2}$ sq. ft.

$\qquad\qquad\qquad\qquad = \frac{1}{3} + 1$ sq. ft.

$\qquad\qquad\qquad\qquad = \frac{4}{3}$ sq. ft.

Volume of 36-ft. run $= \dfrac{4}{\cancel{3}} \times \overset{12}{\cancel{36}}$ cu. ft.

$\qquad\qquad\qquad\qquad = 48$ cu. ft.

(b) Weight of 36-ft. run $= 48 \times 200$ lb.

$\qquad\qquad\qquad\qquad = 9600$ lb.

$$= \frac{\overset{30}{\cancel{\underset{9600}{\cancel{240}}}}}{\underset{\underset{7}{\cancel{56}}}{\cancel{2240}}} \text{ tons}$$

$\qquad\qquad\qquad\qquad = 4\frac{2}{7}$ tons.

It can be shown that the volume of a sphere (a ball-shaped object) is $\dfrac{4\pi}{3} \times r^3$, where r is its radius.

Note. In this chapter general rules have been given for calculating areas and volumes in non-mathematical language. Thus while the man in the street would most likely say that the area of a rectangle is length × breadth, the purist would insist on using some such words as, "The number of square units in the area of a rectangle is equal to the number of units in the length multiplied by the number of units in the breadth." Similarly for the other rules given for areas, and volumes.

EXERCISE 21

1. Find the areas of the following figures (take $\pi = \dfrac{22}{7}$):

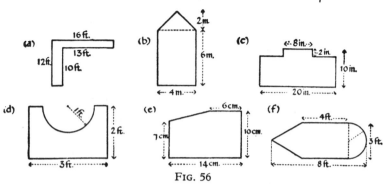

Fig. 56

2. A girder is 40 ft. long and its cross-section is illustrated in Fig. 57. What is its weight in tons if it is made of metal weighing 360 lb. per cubic foot?

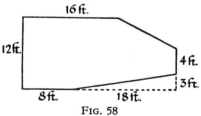

Fig. 57

3. Find the total area of Fig. 58.

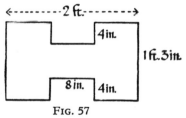

Fig. 58

4. What is the area of a front face of the wall in Fig. 59? If it extends to 24 ft. under the bridge what volume of stonework

is required to construct the two supporting walls, assumed solid throughout?

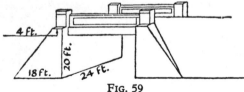

FIG. 59

5. The circles in Fig. 60 represent holes $1\frac{3}{4}$ in. in diameter, punched out of a plate.

(a) Find the surface area in sq. in. of the remainder of the plate (nearest sq. in.).

(b) Find the volume in cu. in. if the plate is $\frac{1}{2}$ in. thick (nearest cu. in.).

Take $\pi = \dfrac{22}{7}$.

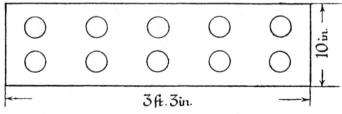

FIG. 60

6. The ends are semicircular, and the holes (radii 1 in.) are punched out (Fig. 61). Find the surface area in sq. in. of the remainder of the plate.

Take $\pi = \dfrac{22}{7}$, and give the answer to the nearest sq. in.

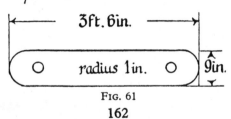

FIG. 61

162

CHAPTER XI

APPLIED GEOMETRY (I)

OUR AIM

GEOMETRY, as we have seen, started by being practical; it had to do with the everyday things of life. The Greeks intellectualized it, and so theoretical rather than practical geometry became the heritage—often the sad heritage— of our schoolchildren until quite recently. Nowadays the emphasis is returning to the practical, and it is chiefly from that view-point that we shall here consider geometry. That is not to say that we shall occupy ourselves with the mechanics of drawing parallels, bisecting angles, and other such things dear to the schoolbook. Instead, we shall try to find out how geometry is applied in the arts and crafts; and, later, we shall see how the 'measuring' geometry of the practical instruments has become the 'calculating' geometry which we call numerical trigonometry.

Mention has already been made of geometry in architecture, and the reader has been invited to look for the commoner geometrical shapes in the workshop, in the factory, and, in fact, everywhere round him.

SOLID INTO FLAT

The formal geometry of the last two thousand years has been largely occupied with lines, angles, triangles, etc., depicted on a flat surface. On the principle of abstraction all problems were reduced to the drawing or study of lines which could be reproduced on a flat sheet of paper, on a board, or, in olden times, on sand.

But the objects among which "we live and move and have our being" are for the most part solid; they have bulk. And so we have the apparent paradox that our geometry so to speak flattens out everything. Of course, it will be argued—what about a photograph, or the image on a cinema screen? They are flat. That is certainly so, yet they give the impression of depth as well as of length and of breadth. This is more obvious in the case of a painting. Here the art of the painter produces an idea of depth in a much more real manner than the blacks and the whites and the greys of a photograph.

This, then, will be our first line of investigation: how is it possible to reproduce on a flat surface in an adequate manner an object, or group of objects, which we know to be solid? Distortion of some kind, or omissions, there must obviously be. We shall try to discover the underlying principles.

SCALE-DRAWING

We shall assume that the reader is familiar with the elementary principles of scale-drawing. When a plan is drawn to a scale of 1 in. = 1 ft. we know, for instance, that a length of $6\frac{1}{2}$ in. on the plan denotes an actual length of $6\frac{1}{2}$ ft., or 6 ft. 6 in.; and a room 15 ft. 9 in. long would have a length of $15\frac{3}{4}$ in. on the plan.

We expect the plan of a house to be an exact copy, on a smaller scale, of the actual shape and dimensions of the house. The discussion of what ensures that the plan shall be an exact copy of the original we shall have to postpone for a while; our intuition, if you like, tells us what to expect in the plan when we know the house, or what to expect in the house when we have seen the plan.

Map-drawing is the same thing carried to a further degree, though here the problem is complicated by the fact that a curved surface (the surface of the earth) has

164

to be copied on a flat surface when a globe is not used. Hence the various 'projections'—Mercator's, conical, and so forth—on which maps of a country or a continent or the world are drawn.

ROUGH SKETCHES

For simplicity we shall start with a rectangular box, or *cuboid*, to give it its geometrical name. We shall suppose that it is 5 ft. long, 4 ft. wide, and 3 ft. deep.

FIG. 62   FIG. 63

A rough sketch of the block is easily made in the manner illustrated in Figs. 62 and 63. In the former, two rectangles, roughly 5 ft. by 3 ft. (scaled, of course), have been drawn with their sides parallel in the positions shown in the sketch. Corresponding corners have been joined in Fig. 63, and to complete the sketch broken lines have been left in to show the edges which would not appear in a side view.

PLANS AND ELEVATIONS

The draughtsman, with some such idea as is given by the sketch in Fig. 63 in mind, would draw a plan (the ground base) and one or two elevations (front, and side or end) of the cuboid from which a workman could read all that he required for its construction.

In this case all lengths are drawn accurately to scale; dimension lines are also affixed as in Fig. 64, and (or) a scale is given by means of which the lengths of lines taken

165

from the drawings can be translated into terms of actual lengths.

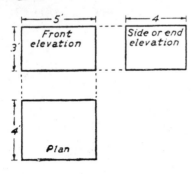

FIG. 64

The broken lines connect the ends of corresponding edges. If the front elevation and the end elevation were hinged and folded through a right angle out of the plane of the paper and then the end elevation through another right angle they could be fitted together to form with the plan three faces of the cuboid.

ISOMETRIC PROJECTIONS

A fairly good idea of the appearance of the solid is given by means of an isometric projection (from the Greek words *isos*, equal, and *metron*, measure).

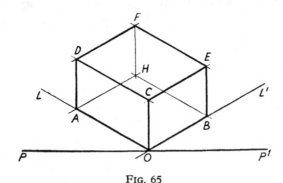

FIG. 65

A line *POP'* is first ruled parallel to the bottom of the page. Lines *OL, OL'* are then ruled, making angles

166

of 30° with *OP* and *OP'*. Scaled lengths *OA* and *OB* are then measured along *OL* and *OL'* equal to the length and breadth of the solid (here 5 ft. and 4 ft.). With a set-square verticals are drawn through *A*, *O*, and *B*. The scaled length *OC* is then cut off equal to the depth of the solid. With a set-square, *CD* is drawn parallel to *OA*, *CE* to *OB*, *DF* to *CE*, and *EF* to *CD*. (Of course the whole thing may also be sketched free-hand.) If the 'hidden' edges are also to be shown, parallels to *OB* and *OA* through *A* and *B* respectively are drawn and *FH* joined. (As a check on the accuracy of the construction, *FH* should be vertical.) Finally, the 'showing' edges are more heavily ruled.

PERSPECTIVE

The isometric projection shown in Fig. 65 is easily drawn (and it is a most interesting occupation), but it does not look quite like the real thing. If a person stands on a footbridge over a railway consisting of straight sets of rails and looks up and down the lines they appear to the eye to meet in the distance. That they do not actually meet is obvious, yet they *appear* to do so. This is due to the property known as *perspective*. It is the way in which one of nature's most wonderful organs—the eye—receives and harmonizes its impressions.

Look at a rectangular box, or cuboid, from above and at a distance from it. Faces which one knows to be rectangles appear to be trapezia or just quadrilaterals; lines which are parallel appear as if they would meet when prolonged sufficiently far to the left or the right. But vertical lines always look as if they are vertical.

Putting aside theory, we shall say outright that the horizontal parallels appear to meet in the distance at two points which are called the *vanishing points*; but vertical lines always appear as verticals.

We can utilize this property to make a drawing of a solid which looks more 'real' than an isometric projection; yet it is constructed on much the same principle.

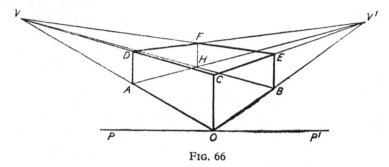

FIG. 66

A straight line *POP'* is drawn, as for isometric projection, parallel to the bottom of the page, and two vanishing points *V* and *V'*, at the same level, are marked.

Join *OV*, *OV'*. Mark off *OA*, *OB*, *OC* (perpendicular to *POP'*) as before to represent the length, breadth, and depth of the cuboid. Join *CV*, *CV'*. Draw the verticals *AD*, *BE*. Join *EV*, *DV'*, meeting at *F*. If the 'hidden' edges are required *BV*, *AV'*, and *HF* are also joined.

If all lines except the edges of the cuboid are now erased the resulting figure bears an astonishing resemblance to the picture of the cuboid as seen by our eyes. (Here again the work may be done free-hand.)

More Difficult Drawing. The reader, after trying his hand and eye on the simple form of the cuboid, may like to reproduce the following drawings of a shed having a door in one end and a window in one side (Fig. 67).

It is hoped that this brief treatment will be of assistance not only to those whose work involves the making and reading of dimensioned drawings, but also to those who wish to deepen their sense of appreciation of art, be it in drawing, painting, nature, or construction.

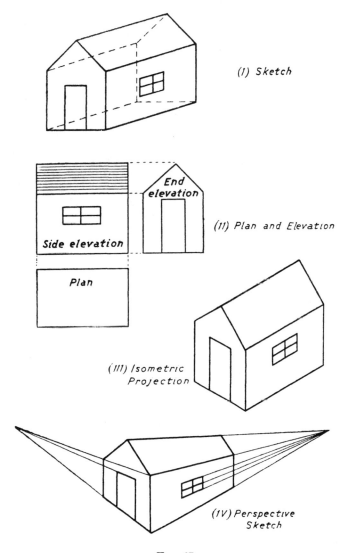

(I) Sketch

End elevation

Side elevation

(II) Plan and Elevation

Plan

(III) Isometric Projection

(IV) Perspective Sketch

Fig. 67

The principle of perspective is especially important. It is basic in all forms of sketching, and, indeed, in all art. Many a sketch has been spoiled because the artist has not paid sufficient attention to this.

It is suggested that the reader experiment with perspective drawing; with a little practice even the most non-artistic can produce satisfactory results. It will be better to start with ruler and set-square for verticals; when facility is gained with this, free-hand work should then be attempted.

CHAPTER XII

APPLIED GEOMETRY (II)

GEOMETRICAL COPYING

How often have we not heard the remark made about the photograph of a friend "That is a speaking likeness of him," or words to that effect? And if they were shown a distorted picture of, say, St Paul's Cathedral or Westminster Abbey nine hundred and ninety-nine people out of a thousand would be quick to point out that there was something wrong about it somewhere or somehow—and the odd man out would probably have something wrong with his vision.

What is it that makes one photograph, or plan, or drawing, or model an exact copy, on a larger scale (as in the case of a cinema picture) or on a smaller scale (as in the case of a photograph, drawing, or plan), of another? This is where geometry—and in this case geometrical theory—can help us.

At the start of our investigation let us remind ourselves that two things are basic in fixing position: measurement of distance, or length of lines; and measurement of angles, or separation between lines.

CONDITIONS OF SIMILARITY

We expect a plan of a house—strictly speaking, the plan of the floor of a house—to reproduce faithfully the exact angles between the walls, etc., and, in accordance with the scale used, the exact lengths and breadths of the rooms, the thickness of the walls, fireplaces, doors, windows, and so forth. What does this mean from the point

171

of view of geometry? We say in geometry that the plan is *similar* to the original of which it is a copy: each angle in the plan is equal to the corresponding angle in the original; and each length in the original is reduced in the same ratio—depending upon the scale used—in the corresponding length of the plan.

We have here the essential conditions for geometrical similarity.

If two figures are similar their corresponding angles are equal and their corresponding sides are in the same ratio— that is, they are proportional.

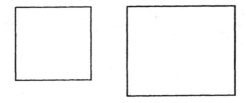

FIG. 68

Both conditions are generally necessary. Compare a square and a rectangle (Fig. 68). Their corresponding angles are equal, for they are all right angles. But the figures are clearly not of similar shape. Why? Their corresponding sides are not proportional.

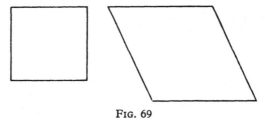

FIG. 69

Now compare a square and a rhombus (Fig. 69). The sides of a square are all equal; the sides of a rhombus are

172

also all equal. So the corresponding sides of the square and the rhombus are proportional. Yet the figures are clearly not of similar shape. Why? Their corresponding angles are not equal, for the angles of a rhombus are not right angles.

Of course, all squares are similar, all equilateral triangles are similar; so are all regular hexagons, for they fulfil the conditions of equal corresponding angles and proportional corresponding sides.

The discussion of the similarity of figures which include curves in their structure is rather beyond the scope of this work. But it is evident that all circles are similar figures, and that all portions of circles bounded by radii which include equal angles are similar.

AN IMPORTANT EXCEPTION

In the triangle *ABC*, *DE* has been drawn parallel to *BC*. There is no angle between *DE* and *BC*, and so *DE* and *BC* make equal angles with *AB* and also with *AC*. Thus the corresponding angles of the triangles *ADE* and *ABC* are equal. We say that the triangles are *equiangular*. Now it can be shown by measurement, or appreciated

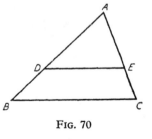

Fig. 70

by inspection, or proved by the methods of theoretical geometry, that the corresponding sides of the triangles *ADE* and *ABC* are in the same ratio—that is, are proportional.

Thus if two triangles are equiangular their corresponding sides are proportional. It can similarly be shown that if the corresponding sides of two triangles are proportional their corresponding angles are equal.

173

Now we have seen that in general two figures built from straight lines are similar if their corresponding angles are equal *and* their corresponding sides are proportional. In the case of the triangle it will be noticed from the work which has just been done that *one* of these conditions is sufficient; for if *either* condition holds the other also necessarily holds.

Thus two triangles are similar *either* if they are equiangular *or* if their corresponding sides are proportional. (This excursion into theory is necessary as a preparation for the introduction later of trigonometry.)

Note. It will be remembered that two triangles are equiangular if two angles of the one are equal to the corresponding two angles of the other; for the third angles must be equal, by the angle-sum property of a triangle (see p. 139).

CONGRUENCE

If two similar triangles have also a side of the one equal to a corresponding side of the other (in the ratio of 1 : 1, if you like) the triangles must be identically equal, angles to angles and sides to sides. They are said to be *congruent*, or *equal in all respects*. We shall here just give a list of the conditions under which two triangles are congruent.

(i) Three sides of the one equal to the corresponding three sides of the other (Fig. 71).

FIG. 71

(ii) Two sides of the one and the angle between them equal to the corresponding sides and angle of the other (Fig. 72).

174

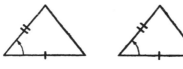

FIG. 72

(iii) Two angles and any side of the one equal to the corresponding angles and side of the other (Fig. 73).

FIG. 73

(iv) In the special case of right-angled triangles—that is, triangles having one angle a right angle—the equality of any two pairs of corresponding sides is sufficient.

It might be thought from the above that the discussion of the congruence of triangles is purely academic. It is of fundamental importance from this point of view, but it also has very practical applications. For instance, one and only one framework can be made from three laths or pieces of metal which are of given lengths. To put it in a different way, a single brace across a corner of a loose framework is sufficient to 'stiffen' that corner. (Is a further limitation of the facts necessary here?) The reader may like to trace similar results for the other cases of congruence.

SYMMETRY

Most people make use of the ideas of symmetry in their daily lives without realizing that they are thereby employing a concept which is geometrical in origin. We know intuitively what it is that makes many shapes and forms and arrangements pleasing to the eye. Often it is because they are nicely balanced. The housewife usually

175

arranges ornaments or pictures in a balanced pattern; windows and many articles of furniture would look strange if one side differed from the other in shape. And who ever saw a man clean-shaven on one side of his face and "bearded like the pard" on the other? Of course,

A DESIGN FOR LINOLEUM

there are other things which attract for the very opposite reason. Take, for instance, a woman's hat or a stretch of countryside or a good picture. But there are some things that definitely owe their attractive appearance to the fact that they are well balanced, where everything on the one side is balanced, or faced, by an exactly similar (in fact, congruent) thing on the other side.

The geometrical word which expresses this idea of balance is *symmetry*; and the object so described is said to be *symmetrical*.

176

LINE SYMMETRY

Take a square or rectangular sheet of paper (newspaper will do very nicely), and fold it so that an edge falls exactly upon the opposite edge. Continue the folding *ad lib.*, taking care always that the folding is done in such a way that an edge always falls upon an opposite edge. (In the case of a square sheet of paper an edge may also be folded to fall on the edge next to it.) Finally a 'nick,' or 'nicks,' of any shape may be cut out of any of the edges appearing after the last folding.

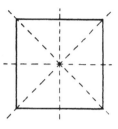

FIG. 74

If the paper is now opened out flat again a pattern of creases and cut-outs will appear. The oftener the paper has been folded and the more in number and the more elaborate in shape the 'nicks' made, the more intricate will be the resulting design. But however elaborate the design, it will be found that it is symmetrical about the first crease mark which was made. It may also be symmetrical with respect to other crease marks.

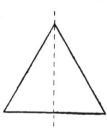

FIG. 75

The patterns so formed are said to have *line symmetry.* The lines which divide the pattern into two exactly like portions are called *axes of symmetry.*

Of course, there is no need to cut paper to demonstrate line symmetry. It is clear, by folding, that a square has line symmetry about the straight lines joining the middle points of its opposite sides, and also about its *diagonals*—that is, the straight lines joining opposite corners (Fig. 74).

M   177

Similarly, an isosceles triangle has line symmetry about the straight line which divides the angle between the two equal sides into equal parts (Fig. 75).

GATEWAY OF BATTLE ABBEY

The Gateway is *almost* symmetrical about a centre line. Why 'almost'?

The façades of many of our most famous buildings also have line symmetry. Most people are familiar with the west end of St Paul's Cathedral and of Westminster Abbey. Except for some small deviations they are good examples of line symmetry.

USING LINE SYMMETRY

As, when there is line symmetry, one side of a figure can be folded over the line or axis of symmetry to fall exactly upon the other side, it follows that the line joining a point to its corresponding point on the other side of the axis of symmetry is *bisected* (that is, divided into two equal parts) at right angles by the axis of symmetry.

178

Thus in Fig. 76 $FL = LB$, $EM = MC$, and AD is perpendicular to FB and to EC.

We can use this principle to carry out a design of which one side is given, the other having to be completed with reference to a given axis of symmetry.

In Fig. 77 the half-drawing $LABCDEFM$ is given. It is required to complete it with respect to LM as axis of symmetry.

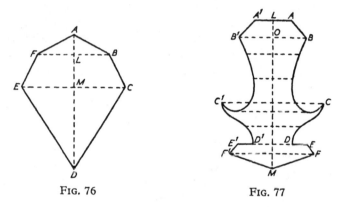

FIG. 76   FIG. 77

Perpendiculars are first drawn from A, B, C, E, F to LM (the axis of symmetry) and also from convenient intermediate points on the curved portions. These perpendiculars are produced beyond LM to an equal length, and the points A', B', C', etc., so obtained are joined by straight lines or curves as the case may be.

Apart from its practical value, the reader will find that making the 'double' of a design in this way is quite a pleasant way of passing the time.

OTHER KINDS OF SYMMETRY

Corresponding to line symmetry in a plane we have *plane symmetry* in a solid (or hollow) body. The reader will readily think of examples of buildings, vessels, etc., which can be divided into two identically equal halves by a plane.

Then there is *point symmetry*, which can be observed in flat, solid, and hollow objects. In point symmetry everything is balanced about a point called the *centre of symmetry*. Each point in the design or object has its exact counterpart, which may be found by joining the point to the centre of symmetry and producing the join to an equal length on the other side of the centre of symmetry. The most elementary examples of point symmetry are circles and spheres. The centre of the circle or sphere is the centre of symmetry. An ellipse also has point symmetry about its centre as well as line symmetry about its axes.

Lastly, there is *radial symmetry*. Examine Fig. 78. If the whole design is rotated through a third of a whole

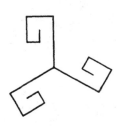

revolution about the point where the three lines meet, each branch takes up the exact position vacated by the next branch. This kind of symmetry is called radial symmetry; the point about which the rotation takes place is called the centre of symmetry.

FIG. 78

We see radial symmetry in the petals of well-formed flowers— daisies, anemones, and sunflowers are a few examples —and in starfish on the seashore.

To test his knowledge of the principles of symmetry the reader may care to investigate the capital letters of the alphabet with a view to discovering what kind, or kinds, of symmetry, if any, each possesses.

CHAPTER XIII

APPLIED GEOMETRY (III)

APPLICATIONS OF STANDARD PROPERTIES

PURSUING our policy of considering the application of geometry to everyday problems, we shall in this chapter consider how some of the standard properties of the commoner geometrical forms have been turned to practical use. For the present we shall not attempt to prove the properties which we quote. In some cases their truth is almost self-evident.

PYTHAGORAS'S THEOREM

Among the geometrical truths which have been known for well over two thousand years Pythagoras's Theorem

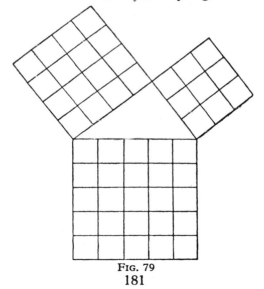

FIG. 79

is probably one of the best known generally. Ahmes (about 1650 B.C.), whom we mentioned in connexion with fractions at p. 17, knew that a triangle whose sides were 3, 4, and 5 units long was right-angled, the right angle being enclosed by the two shorter sides.

Let us check it. Draw, on squared paper if possible, three squares whose sides are 3 in., 4 in., and 5 in. long, and fit them together as in Fig. 79. Measure the top angle of the triangle with a protractor or set-square. It is 90°—that is, a right angle.

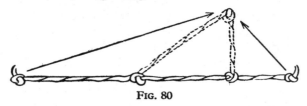

FIG. 80

This principle was used many thousands of years ago by the Egyptian surveyors (or 'rope-stretchers,' as they were called) to draw a right angle. The river Nile regularly overflowed its banks, and so their services were often required. They used a rope knotted at lengths of 3, 4, and 5 units to construct a right angle (Fig. 80).

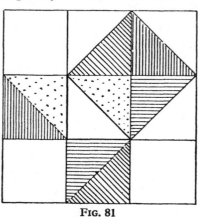

FIG. 81

They also knew that in a right-angled triangle with two equal sides the square on the side opposite the right angle was equal to the sum of the squares on the other sides. Can you see it for yourself in this section of a floor covered with square tiles (Fig. 81)?

It was the Greek Pythagoras (sixth century B.C.) who first discovered that it is a peculiarity of *all* right-angled triangles that the square on the side opposite the right angle is equal to the sum of the squares on the sides enclosing the right angle. He was so excited at his discovery that he promptly sacrificed an ox!

Let us check this on the 3, 4, 5 triangle of Fig. 79.

$3^2 = 9$, $4^2 = 16$, so $3^2 + 4^2 = 9 + 16 = 25$, and $5^2 = 25$.

Other right-angled triangles have sides of 5, 12, 13, and 7, 24, 25 units; and, of course, multiples and sub-multiples of the sides of these triangles give right-angled triangles—*e.g.*, 6, 8, 10, and 10, 24, 26 units.

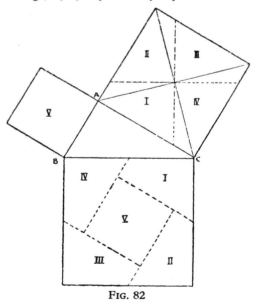

FIG. 82

Pythagoras, and many others, have discovered geometrical proofs for this. You may like to check its truth

183

for yourself by cutting out and fitting together a copy
of the jig-saw illustrated in Fig. 82.

ABC is a right-angled triangle with the right angle
at *A*. Draw the squares on the sides. Get the centre of
the square drawn on *AC* by finding where the lines
joining opposite corners meet. Divide this square by
lines through the centre parallel and perpendicular to *BC*.

Cut out the upper I, II, III, IV (the inner boundaries
are the dotted lines), and V, and show that they fit into
the square on *BC* as shown in the diagram. This is a
real jig-saw puzzle!

We have thus checked what is usually known as
Pythagoras's Theorem, or the Rule of Pythagoras:

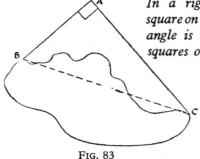

*In a right-angled triangle the
square on the side opposite the right
angle is equal to the sum of the
squares on the other two sides.*

The importance of
this for us is that it
enables us to calculate
the length of the third
side of a right-angled

FIG. 83

triangle if we know
the lengths of two of its sides. Take this example:

It is required to find the distance between the two
points *B* and *C* on a lake. A point *A* is taken on the
land, its position being such that *AB* is at right angles
to *AC*. *AB* and *AC* are measured and found to be
23 yd. and 47 yd. respectively.

The square on $AB = 23^2 = 529$;
and the square on $AC = 47^2 = 2209$.

But by Pythagoras's Theorem, since angle *A* is a right
angle, the square on *BC* is equal to the sum of the
squares on *AB* and *AC*—that is—529 + 2209, or 2738.

184

Now $50^2 = 2500$ (too small), by actual multiplication,
$55^2 = 3025$ (too large),
$51^2 = 2601$ (too small),
$53^2 = 2809$ (too large),
$52^2 = 2704$ (fairly near).

So *BC* is roughly 52 yd. long.

Of course, as we shall see later, we can obtain $\sqrt{2738}$ by more accurate methods.

EXERCISE 22

1. The sides of a rectangular plot are 9 yd. and 12 yd. long. What is the distance between the opposite corners?

2. When flying along a certain track to the east of north for 25 miles a pilot notices that he has flown 24 miles to the north of his taking-off point. How far east has he flown?

3. What is the distance between the centres of the bolt-holes in Fig. 84?

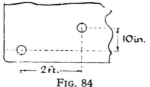

FIG. 84   FIG. 85

4. Find the length of the rafter to meet the specification given in Fig. 85.

ARITHMETIC AND GEOMETRY DO NOT QUITE AGREE HERE

If a right-angled triangle is drawn in which the sides including the right angle are 3 in. and 5 in. long then by Pythagoras's Theorem the square on the side opposite the right angle $= 3^2 + 5^2 = 9 + 25 = 34$ sq. in. And so the length of the side opposite the right angle $= \sqrt{34}$ in.

185

If the construction is actually performed a line which is $\sqrt{34}$ in. long can be drawn and measured. But the square root of 34 is not an exact number; it lies between 5 and 6, but no number can be found whose square is exactly 34. Of course, there is a way of extracting the square root of 34 by arithmetical methods, but the result has to be corrected, or rounded off, to an arbitrarily chosen number of places of decimals. The square root of 34 is not even a recurring decimal (see p. 72).

Here, then, is another way in which our decimal notation does not exactly do all we want it to do (and the same applies also to fractions). The alert reader will at this stage say, "But at any rate we can calculate the square root of 34 to more places of decimals than anyone could possibly measure accurately." And he would be correct; and there we shall leave it.

THE ISOSCELES TRIANGLE

An isosceles triangle is a triangle with two equal sides. It can be proved that the angles at the base of an isosceles

FIG. 86

triangle are equal—in other words, if two sides of a triangle are equal the angles opposite them are equal. In Fig. 86 if $AB = AC$

then angle B = angle C.

If the legs of a step-ladder are equal then each is equally inclined to the ground. In house-building rafters of equal length which meet at a ridge are at equal slopes with the horizontal if their lower ends are at the same level.

186

PARALLELOGRAMS, RECTANGLES, AND SQUARES

For definitions see pp. 140-141. These three figures have in common the properties that their opposite sides are equal (and parallel) and that their diagonals bisect each other.

The parallelogram is used in wall-brackets (Fig. 35). In whatever position the bracket is clamped the supported tray is always horizontal; for the stay supporting it must always be parallel to the wall—that is, vertical.

Rectangles and squares also have the important property that their diagonals are equal. As we have already seen, a carpenter can tell whether a window or door-frame is 'squared' by ensuring that the opposite sides are equal, and also the diagonals; $AB = DC$, $AD = BC$, $AC = BD$ in Fig. 87.

FIG. 87

The centre of a rectangular or square board (and, ladies, a piece of linen or other material) is found by finding the point of intersection of the diagonals.

THE TANGENTS TO A CIRCLE FROM A POINT OUTSIDE THE CIRCLE

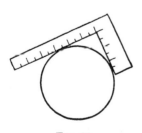

FIG. 88

It can be proved that if two tangents are drawn to a circle from a point outside it then the lengths of the tangents measured from the external point to the points where the tangents touch the circle are equal.

A round object can be tested to find whether it is a perfect circle or a perfect sphere (a ball) by placing a carpenter's square in the position shown in Fig. 88. If the 'square'

187

meets the round object always at equal distances from the inner corner of the 'square,' the object is a perfect circle or sphere.

If a string or belt in passing from a point A outside a cir-

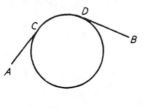

FIG. 89

cular pulley wheel to another point B outside it is tightly stretched its path will follow the tangents from A and B to the circle and the portion of the circumference between the points of contact of the tangents (Fig. 89).

The 'lie' of a tight belting connecting two pulleys (direct or cross-belting) is such that the straight portions of the belt touch the outer rims of the pulleys.

THE ANGLE IN A SEMICIRCLE

A *semicircle* is half a circle. A *hemisphere* is half a sphere. A semicircle is bounded by a diameter and the part of the circumference of the circle cut off by the diameter. A hemisphere is bounded by a plane through the centre of the sphere and the portion of the surface of the sphere cut off by that plane.

The *angle in a semicircle* is the angle between two straight lines which connect any point on the circumference of the semicircle to the ends of the diameter (Fig. 90). It can be proved that this angle is always a right angle, wherever the point is taken on the circumference.

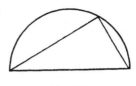

FIG. 90

The practical man can check whether a rounded hollow is a perfect hemisphere by placing a 'square' so that its two outside edges rest always at the ends of a diameter of the top section, supposed circular. (The

188

previous test will check
this.) If the corner of the
'square' just meets the surface of the hollow for all
such positions of the
'square' (Fig. 91) the hollow is a perfect hemisphere.

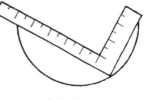

FIG. 91

MORE ABOUT SIMILAR FIGURES

Finding an area is often a matter of calculating length multiplied by breadth. So if two figures are geometrically similar it should not surprise the reader to know that the ratio of their areas is equal not to the ratio of their corresponding sides, but to the ratio of the squares on their corresponding sides; for the length and breadth are each increased or decreased in the same ratio.

Thus, in scale drawing if the length scale used is 1 in. = 10 ft. the scale for areas will be 1 sq. in. = 100 sq. ft.

Similarly, when the volume of a model is compared with that of the original the ratio of the volumes is the ratio of the cubes of the lengths compared.

BISECTING AN ANGLE

Here is a way of bisecting an angle—say, an angle

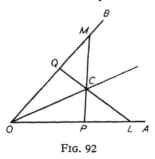

FIG. 92

AOB—using a straight-edge only. Equal lengths OL and OM are marked on OA and OB (Fig. 92). Another pair of equal lengths, OP and OQ, are marked on OA and OB. Let LQ and MP meet at C. Join OC. Then OC bisects the angle AOB.

This construction has two

189

especially interesting features. First, a compass is not necessary (nor, indeed, is a graduated ruler-scale, for the equal lengths can be marked on any straight edge); secondly, the proof of the validity of the construction (we shall not attempt it at this stage) involves each of the three cases of congruence.

TAILPIECE

These practical applications of theoretical geometry are offered as a few examples of what can be done with this powerful tool. Some readers may perhaps wonder when they look back on the discipline which geometry was to them in their schooldays, "Was it worth it?" The answer to this question must, of course, vary in the case of each individual. But we shall see later, when we touch on the theme of 'arguments,' that the value of geometry is not to be assessed merely from the point of view of its practical applications; its utility as a trainer of the mind in orderly thought will also have to be considered.

CHAPTER XIV

APPLIED GEOMETRY (IV)

NUMERICAL TRIGONOMETRY

WHEN a survey has to be made of a part of a country or of the heavens for map-making angles have to be measured. The land surveyor measures distances and angles; the astronomer measures time and angles. The gunner also has to make use of angles; he does it by adjusting a 'sight.'

Now direct measurements of length are only infrequently possible with any accuracy. Trees, rivers, houses, and the irregularity of the land all get in the way. Actually in the making of the Ordnance Survey maps of this country only a few direct measurements of length have to be made on flat pieces of land such as are to be found on Salisbury Plain, and some of these measurements are made only for checking purposes. The explanation of this is as follows.

We have seen that a triangle can be drawn if one of its sides and two of its angles are known. So, starting from a base-line AB, if C is another point, not on AB, and the angles CAB, CBA can be measured, the triangle CAB can be drawn and the distances CA and CB measured from it. Then from CB and CA as base-lines, if the angles to fresh points D or E can be measured, the triangles CDB, CEA can be drawn (Fig. 93). In this

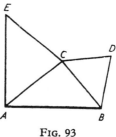

FIG. 93

manner the whole country can be completely covered with triangles.

Thus mapping out a country by 'triangulation' really means, after the first length has been accurately determined, measuring angles. It is from this that the word *trigonometry* is derived. *Trigonon* is the Greek word for a 'triangle,' *metria* a Greek word-ending for 'measurement.'

Now, even with extremely accurate measurements, scale drawings would be an inadequate way of tackling the mapping of a country, or the heavens, however carefully the drawings were made. So instead of this was developed—first for the purposes of astronomy, nearly two thousand years ago, and later as a convenient tool for land-surveying—the science or art of trigonometry—the replacement of scale drawing by calculation.

There are two approaches to trigonometry: the theoretical and the practical. We shall deal briefly with each of them. It may be that the theoretical may appeal to one class of reader, the practical to another.

THE THEORETICAL APPROACH

We have seen (p. 173) that the lengths of corresponding sides of equiangular triangles are proportional to each other—that is, in the same ratio.

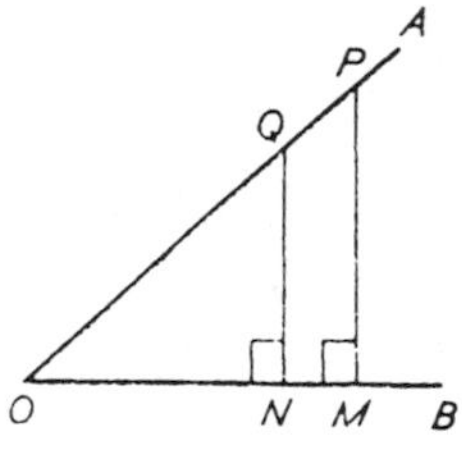

FIG. 94

Take a point P on one arm OA of an angle AOB, and draw a perpendicular PM to the other arm OB (Fig. 94). (The square mark ⌐ is nowadays generally used to denote a right angle.)

From any other point Q on OA draw QN also perpendicular to OB.

Then the triangles QON, POM are equiangular. (Why?) And so

$$\frac{QN}{ON} = \frac{PM}{OM}.$$

Now we deliberately said "from *any* other point Q on OA." So wherever Q is taken on one arm of $A\hat{O}B$ the value of the ratio $\dfrac{QN}{ON}$ is always the same for that angle. It is also clear that as the size of the angle is altered so the value of the ratio also changes. The ratio is therefore a property of the angle, as it does not depend upon where any particular perpendicular is set up.

This ratio is called the *tangent* of the angle, and it is written:

$$\tan A\hat{O}B.$$

The tan ratio of any particular angle—say, tan 23°— may be obtained by making an angle of 23°, drawing a perpendicular from one arm of the angle to the other arm, measuring the necessary lengths, and calculating the value of the ratio by ordinary division. Of course, in most cases it will have to be 'rounded off' to any given number of places of decimals or number of figures—say, two, or three, or four. In practice, however, these ratios are read from tables which have been worked out by means of more advanced mathematics than we need to consider.

THE PRACTICAL APPROACH

The more practically minded may better appreciate the following line of argument.

(*a*) Draw a triangle ABC in which $AB = 4$ in., $BC = 3$ in., angle $B = 90°$. On AB mark off $AB_1 = B_1B_2 = B_2B_3 = 1$ in. Draw B_1C_1, B_2C_2, B_3C_3, each perpendicular to AB. Make the necessary measurements, and fill in the following table (divisions to two places of decimals):

$\dfrac{BC}{AB}$	$\dfrac{B_3C_3}{AB_3}$	$\dfrac{B_2C_2}{AB_2}$	$\dfrac{B_1C_1}{AB_1}$

Now take *any* point P on AB (or AB produced), and draw PQ perpendicular to AB (or AB produced)

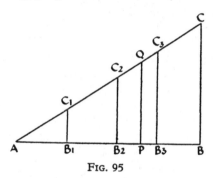

FIG. 95

to meet AC (or AC produced) at Q. Measure AP and PQ, and fill in as above:

$$\frac{PQ}{AP} =$$

From this you will see that, wherever you take the point P and draw PQ perpendicular to AB to meet AC at Q,

the value of the ratio $\dfrac{PQ}{AP}$ is the same.

(b) Draw any right-angled triangle XYZ (base XY, YZ perpendicular to XY), and show, as above, that

194

if L is any point on XY, and LM, drawn perpendicular to XY, meets XZ at M, then:

$\dfrac{LM}{XL}$ is the same wherever L is taken on XY;

but that the values obtained in (*a*) above are not the same as those obtained here (unless, of course, the triangles happen to be congruent, or of the same shape).

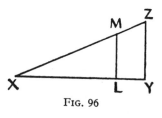

FIG. 96

From this you will see that the values of these ratios depend upon the size of the angle between the lines, and not on the position of the point on one of the lines where the perpendicular to it is drawn.

TWO MORE TRIGONOMETRICAL RATIOS

Referring to Fig. 94 at p. 192 (or to the corresponding Figs. 95 and 96, and making the consequent alterations in the letters used), we can show in the same ways that for any particular angle:

$$\dfrac{QN}{OQ} \text{ has a definite value; and}$$

$$\dfrac{ON}{OQ} \text{ also has a definite value.}$$

$\dfrac{QN}{OQ}$ is called the *sine* of the angle AOB, and is written $\sin A\hat{O}B$; $\dfrac{ON}{OQ}$ is called the *cosine* of the angle AOB, and is written $\cos A\hat{O}B$ (pronounced 'coss' $A\hat{O}B$).

The sine and cosine of any angle may be obtained by drawing and measurement as explained for the tan ratio at p. 193.

195

SUMMARY

Examine the triangle *ABC* of Fig. 97. It has a right angle at *B*. Instead of naming the sides of the triangle

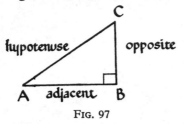

FIG. 97

AB, *BC*, *CA*, we shall call them *adjacent* (to angle *A*), *opposite* (to angle *A*), and *hypotenuse* (the name given to the side opposite the right angle in a right-angled triangle). We can now frame our definitions so that they fit any acute angle. Thus:

$$\text{Tan } \hat{A} = \frac{\text{opposite}}{\text{adjacent}} \quad \text{or} \quad \frac{\text{opp.}}{\text{adj.}}$$

$$\text{Sin } \hat{A} = \frac{\text{opposite}}{\text{hypotenuse}} \quad \text{or} \quad \frac{\text{opp.}}{\text{hyp.}}$$

$$\text{Cos } \hat{A} = \frac{\text{adjacent}}{\text{hypotenuse}} \quad \text{or} \quad \frac{\text{adj.}}{\text{hyp.}}$$

EXERCISE 23

Greek letters are frequently used to denote angles. Some of the commonest in use are: α (*alpha*), β (*beta*), γ (*gamma*), θ (*theta*), φ (*phi*), ψ (*psi*).

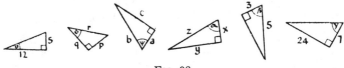

FIG. 98

For the above figures write down: tan θ, sin α, cos β, cos γ, tan ψ, sin θ, cos φ, tan α, sin ψ, cos θ, tan γ, cos α, sin φ, sin γ, tan β, cos ψ, sin β, tan φ. (In numerical cases first calculate the third side, using Pythagoras's Theorem.)

196

APPLIED GEOMETRY

USE OF TRIGONOMETRICAL TABLES

As has already been mentioned we could, if necessary, draw suitable triangles and calculate the sin, cos, and tan of any angle when needed. But this would be tedious, and so tables have been constructed from which the ratios can easily be read. At the end of the book you will find tables of the sin, cos, and tan of any angle from 0° to 90°.

Suppose you have to find sin 58° 12'. (Each degree is divided into 60 minutes, written 60'—and each minute into 60 seconds—60".) Turn to the natural sine table.

The degrees are found in the extreme left-hand column; the minutes at intervals of 6 minutes by looking across the top row. Place a ruler across and under the 58° row, follow with your eye down the column headed 12'; where your eye meets the ruler gives sin 58° 12' = ·8499.

If you have to find sin 58° 16' proceed as above, but now the angle is increased by 4'. Look up this figure (4) in the 'difference' columns on the extreme right of the page; under '4' you will find (in the same row as 58°) the figure 6. Add this to the last figures for sin 58° 12', and we obtain ·8499 + ·0006 or ·8505 for sin 58° 16'.

The same method is used for the tan table.

Thus :   tan 72° 18' = 3·1334;
and   tan 72° 20' = 3·1334
                       +   64
                     = 3·1398.

It can be shown that the cosine of an angle *decreases* as the angle *increases* from 0° to 90°. So the 'difference' for a cosine is *subtracted*, and not added.

Thus :   cos 27° 42' =   ·8854;
but   cos 27° 47' =   ·8854
                       −   7
                     =   ·8847.

197

Conversely, find the angle whose sine is ·7819.

From the tables   ·7815 = sin 51° 24′;

                      4 = difference for 2′;

so, adding for sines, ·7819 = sin 51° 26′.

We find similarly an angle whose tan is given.

To find the angle whose cosine is ·835:

             ·8358 = cos 33° 18′;

                  8 = difference for 5′;

so, subtracting for cosines, ·8350 = cos 33° 23′.

Note. In this reversed process the tables do not always give us the exact figure in the fourth place of decimals; in that case we select the nearest difference figure.

EXERCISE 24

1. Write down: sin 10°, sin 50°, sin 80°, sin 15° 30′, sin 29° 12′, sin 87° 54′, sin 23° 26′, sin 42° 47′, sin 69° 50′.

2. Write down: cos 5°, cos 45°, cos 73°, cos 25° 18′, cos 53° 42′, cos 80° 24′, cos 17° 39′, cos 42° 47′, cos 76° 23′.

3. Write down: tan 14°, tan 45°, tan 60°, tan 20° 6′, tan 49° 36′, tan 78° 48′, tan 13° 25′, tan 46° 20′, tan 75° 45′.

4. Write down the angles whose sines are: ·2250, ·8192, ·3811, ·9317, ·5828, ·9491.

5. Write down the angles whose cosines are: ·9272, ·5000, ·9755, ·3371, ·8, ·265.

6. Write down the angles whose tangents are: ·8391, 1·0, ·5914, 1·6512, 2·5, 4·0.

SOLUTION OF A RIGHT-ANGLED TRIANGLE

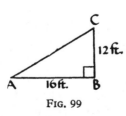

FIG. 99

A triangle is said to be 'solved' when from given sides and/or angles the other sides and/or angles are *calculated*.

(*a*) *Given the two sides containing the right angle*—e.g., *AB* = 16 ft., *BC* = 12 ft.

198

By Pythagoras's Theorem, since angle $B = 90°$,

$$AC^2 = AB^2 + BC^2$$
$$= 16^2 + 12^2$$
$$= 256 + 144 = 400 \text{ sq. ft.}$$
$$\therefore \quad AC = \sqrt{400} = 20 \text{ ft.}$$

Also $\quad \tan A = \dfrac{CB}{AB} = \dfrac{12}{16} = \dfrac{3}{4} = \cdot 7500.$

$\therefore \quad A = 36° \ 52'$ (from the tables).

But $\quad A + B + C = 180°.$

$$\therefore \quad C = 180° - 90° - 36° \ 52'$$
$$= 53° \ 8'.$$

(b) *Given the hypotenuse (the side opposite the right angle) and one side—e.g.*, $AC = 100$ yd., $AB = 40$ yd.

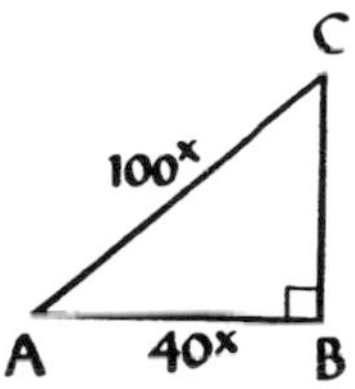

By Pythagoras's Theorem, since angle $B = 90°$,

FIG. 100

$$AB^2 + BC^2 = AC^2.$$
$$\therefore \quad 1600 + BC^2 = 10,000.$$
$$\therefore \quad BC^2 = 10,000 - 1600 = 8400 \text{ sq. yd.}$$
$$\therefore \quad BC = 91 \cdot 7 \text{ yd. (approx.).}$$

Also $\quad \cos A = \dfrac{AB}{AC} = \dfrac{40}{100} = \cdot 4000.$

$\therefore \quad A = 66° \ 25';$

and $\quad C = 180° - 90° - 66° \ 25'$ (angle-sum)
$$= 23° \ 35'.$$

(If we were given BC we should say, $\sin A = \dfrac{BC}{AC} = $ etc.)

(c) *Given the hypotenuse and one angle—e.g.*, $AC = 36$ in., $A = 45° \ 22'$.

Since $A + B + C = 180°$,
$$C = 180° - 90° - 45° \ 22'$$
$$= 44° \ 38'.$$

Also $\dfrac{AB}{AC} = \cos A = \cos 45° 22'.$

$\therefore \quad AB = AC \cos 45° 22'$
$= 36 \times \cdot 7026$
$= 25\cdot3$ in. (on multiplication).

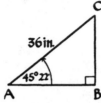

FIG. 101

And $\dfrac{BC}{AC} = \sin A = \sin 45° 22'.$

$\therefore \quad BC = AC \sin 45° 22'$
$= 36 \times \cdot 7116$
$= 25\cdot6$ in.

(d) *Given a side and an angle*—e.g., $BC = 15$ ft., $A = 63° 15'$.

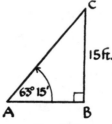

FIG. 102

As above, $C = 180° - 90° - 63° 15'$
$= 26° 45'.$

And $\dfrac{AB}{BC} = \tan C = \tan 26° 45'.$

$\therefore \quad AB = BC \tan 26° 45'$
$= 15 \times \cdot 5040$
$= 7\cdot56$ ft.

200

And, by Pythagoras's Theorem, since $B = 90°$,

$$AC^2 = AB^2 + BC^2$$
$$= 57\cdot15 + 225$$
$$= 282\cdot15 \text{ sq. ft.}$$
$$\therefore \quad AC = \sqrt{282\cdot15} = 16\cdot8 \text{ ft. (approx.).}$$

(Or use $\quad \sin A = \dfrac{BC}{AC}$,

$$\therefore \quad AC = \frac{BC}{\sin A} = \frac{15}{\sin 63° 15'} = \text{etc.}—$$

but this means a more awkward division.)

PROBLEMS ON THE RIGHT-ANGLED TRIANGLE

(*a*) At a distance of a mile from a tower a man lying on the ground has to raise his telescope through an angle of $8°$ from the horizontal to 'sight' the top of the tower.

8°

5280 ft.

FIG. 103

What is the approximate height of the tower?

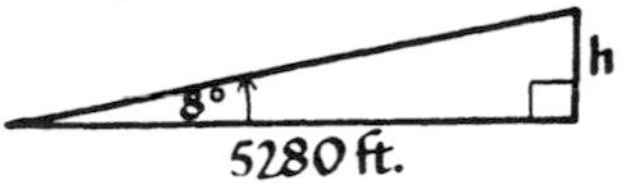

Let the required height be h ft.

Then $\dfrac{h}{5280} = \tan 8°$.

$$\therefore \quad h = 5280 \times \tan 8°$$
$$= 5280 \times \cdot1405$$
$$= 742 \text{ ft. (correct to the nearest foot).}$$

(*b*) A road rises 1 ft. in every 16 ft. measured along its length (not on the level). What is its inclination to the horizontal?

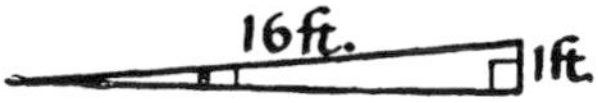

16 ft.

1 ft.

FIG. 104

Let θ be the angle of slope.

Then $\sin \theta = \frac{1}{16}$

$$= \cdot0625 \text{ (on division).}$$
$$\therefore \quad \theta = 3° 35'.$$

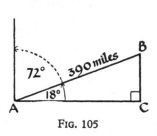

Fig. 105

(c) An aeroplane flies at 260 m.p.h. on a bearing of N. 72° E. How far east of its taking-off point will it be after flying for 1½ hours?

$AB = 1\frac{1}{2} \times 260 = 390$ miles.

Draw BC perpendicular to AC where AC is due east. Then we have to calculate AC.

$\dfrac{AC}{AB} = \cos 18°$ (angle $BAC = 90° - 72° = 18°$.)

So $AC = AB \cos 18°$
$= 390 \times ·9511$
$= 371$ miles (correct to nearest mile).

EXERCISE 25

(Give answers, when not exact, correct to three figures.)

Solve the triangle ABC (a is the side opposite angle A, etc.), given:

1. $a = 4, b = 3, C = 90°$.
2. $b = 13, c = 5, B = 90°$.
3. $C = 90°, c = 25, b = 24$.
4. $b = 16, B = 90°, C = 30°$.
5. $a = 40, C = 90°, B = 65°$.
6. $b = 150, c = 300, A = 90°$.
7. $a = 25, C = 53° 16', A = 90°$.

8. The slope of a railway line is given as 1 in 75. This means that there is a vertical rise of 1 ft. in every 75 ft. in length of the railway. What is the angle of slope to the horizontal?

9. A plane leaves the ground at a steady angle of 12°. Will it clear an obstacle 600 ft. high at a distance of 1000 yd. from the point at which it takes off?

10. How long will the shadow of a wall 20 ft. high be when the angle of inclination of the rays of the sun is 52°?

11. An aeroplane flies 60 miles due west and then 100 miles due north. What is its bearing then from its starting-point? If, instead of flying the 100 miles to the north, it had flown 100

miles due south, what would its bearing then have been from its starting-point?

12. Which is the steeper gradient—a rise of 1 vertical in 10 measured horizontally, or a rise of 1 vertical in 11 measured along the slope?

ADDITIONAL NOTE

The observant reader may have noticed that the sin, cos, and tan of an angle have been defined here only for acute angles.

By a suitable convention of signs these definitions are extended, in the first place, to all angles between 0° and 360° (one complete rotation). Thus:

$$\sin (180° - \theta) = \sin \theta; \quad \cos (180° + \theta) = - \cos \theta;$$
$$\cos (180° - \theta) = - \cos \theta; \quad \tan (180° + \theta) = \tan \theta;$$
$$\tan (180° - \theta) = - \tan \theta; \quad \sin (360° - \theta) = - \sin \theta;$$
$$\sin (180° + \theta) = - \sin \theta; \quad \cos (360° - \theta) = \cos \theta;$$
$$\tan (360° - \theta) = - \tan \theta.$$

For a further discussion of this point see pp. 257-259.

For angles greater than 360° it is evident that the trigonometrical ratios for angles between 0° and 360° are repeated. Thus:

$$1000° = 2 \times 360° + 280°, \text{ and so}$$
$$\sin 1000° = \sin 280°, \cos 1000° = \cos 280°, \text{ and so on.}$$

The trigonometrical ratios are consequently described as *periodic*, the period in each case being 360°, or (p. 158) 2π radians.

PART III

SYMBOLS AND THEIR USES

CHAPTER XV

SYMBOLICAL ALGEBRA

FORMULÆ

ALGEBRA, as we know it, is of comparatively recent date. The Egyptians and the Greeks played about with problems which we to-day would solve by algebra. It is on

THE SIGN OF EQUALITY

From *The Whetstone of Witte*, by Robert Recorde (1557)

record that Hypatia, the first woman mathematician known to history, wrote a commentary on one of the first serious treatises on algebra—that by Diophantus (*c.* A.D. 275). She was later murdered by a fanatical mob—done to death with oyster-shells and finally burnt—but not because she wrote a book on algebra.

THE USE OF THE SYMBOLS + AND −

From *Behede und Lubsche Rechnung,* by Johann Widman
(Leipzig, 1489)

But the greatest writer on algebra before the days of printing was Mohammed ibn Mûsâ al-Khowârizmî (Mohammed, son of Moses, the Khowarezmite). Early in the ninth century A.D. he wrote in Baghdad a book whose title was *al-jebr w'al muqâbalah*. This title has been taken by different people to mean "The Science of Reduction and Cancellation," "Restoration and Reduction," and "The Science of Equations." Whatever its meaning was, a shortened form of it, *Al-jebr*, gave its name to this great branch of mathematics.

The whetstone of witte,

whiche is the seconde parte of
Arithmetike:contaínyng thextrac=
tion of Rootes: The Coßike practise,
with the rule of Equation: and
the woorkes of Surde
Nombers.

From the title-page of the first English algebra book

In those days and for the next few hundred years arithmetic and algebra were treated together in text-books (as is once more happening, a thousand years or so later), and they did not become separated—and then only partially at first—until the invention of the printing press.

The coming of the printing press was followed soon by the invention, and later the standardization, of symbols, and books written in the sixteenth and seventeenth centuries show the progress of algebra from a period in which everything was written in full in words (and what a laborious business it must have been), with only a few abbreviations or contractions, to a time when the work appears in much the same form as that in which we see it to-day.

Without this symbolical algebra—letters instead of some numbers, and signs and symbols instead of long directions written out in full—most of the advances in scientific knowledge would have been almost impossible. Algebra has indeed been a most useful tool in the hands of the engineer and the scientist as well as of the professional mathematician.

At the present time the emphasis in elementary algebra is gradually changing. The algebra of the schools for the last fifty years or so has consisted largely, if the truth be told, of the manipulation of letters, signs, and symbols; to many people it was a species of mental and manual jugglery. Nowadays the tendency is to approach algebra via the formula. We live in a mechanical age; graphs are a commonplace, formulæ are bandied about freely. We even read of statesmen trying to "find a formula."

THE LANGUAGE OF ALGEBRA

Algebra has, as it were, a language of its own to be learned before the subject can be understood or used, and this language has the advantage that it can be used anywhere. Its statements and formulæ are the same in the books of Russia, France, Italy, Spain, and Germany as they are in our books; and in this international language statements can be made more briefly than in any other language.

One of the important features of this language is its use of letters for numbers. In arithmetic we deal with particular numbers, or numbers of things; in algebra we deal with numbers or things in a more general sort of way, and to do this we frequently use letters instead of numbers. For this reason symbolical algebra has been called 'generalized arithmetic.' When we 'translate' statements from arithmetic to algebra we replace words of direction or command, and sometimes adjectives, by symbols; in the result the statement becomes in algebra what is called a *formula*.

§ 62. Rechnerische Lösung der Gleichungen zweiten Grades mit einer Unbekannten. 198

C **1)** Aus geschichtlichen Gründen bezeichnet man die Ergebnisse der Gleichungen meist als Wurzeln der Gleichung.

2) Gleichungen 2. Grades mit 2 Unbekannten, die bei den Anwendungen häufig auftreten, lassen sich sofort auf Gleichungen 2. Grades mit einer Unbekannten zurückführen, wenn die eine Gleichung von erstem Grade ist. Manchmal kann die Lösung auch dadurch erfolgen, daß man einfache Verbindungen der Unbekannten, z. B. $x + y$, $x - y$ gewinnt.

D **1.** a) $x^2 - 8x = 0$;   b) $5x^2 - 3x = 0$;   c) $8x^2 + 12x = 0$.

2. a) $\dfrac{x^2}{2} - \dfrac{3x}{4} = 0$;   b) $\dfrac{4}{5}x^2 + \dfrac{2}{3}x = 0$;   c) $3\frac{1}{4}x^2 + 2\frac{1}{2}x = 0$.

3. a) $x^2 - bx = 0$;   b) $ax^2 - bx = 0$;   c) $a^2x^2 + bcx = 0$.

4. a) $3x^2 - 27 = 0$;   b) $5x^2 = 180$;   c) $\dfrac{x^2}{4} = 16$.

5. a) $\dfrac{4}{5}x^2 - \dfrac{125}{4} = 0$;   b) $\dfrac{x^2}{6} = 150$;   c) $2\frac{1}{4}x^2 = 15\frac{5}{8}$.

6. a) $8x^2 - 62 - 2x^2 = 3x^2 + 13$;   b) $8x^2 - 30 = 6x^2 + 42$.

7. a) $(3x)^2 - 2x^2 = (2x)^2 + 48$;   b) $\left(\dfrac{x}{2}\right)^2 + \left(\dfrac{3x}{2}\right)^2 = 2\frac{3}{4}x^2 - 25$.

THE UNIVERSAL LANGUAGE OF ALGEBRA

A page from a German schoolbook

We have already seen (p. 112) how the calculation of simple interest is made easier by use of the formula:

$$I = \frac{Prt}{100}.$$

It will be remembered that letters placed next to one another without any intervening sign or symbol are regarded as multiplied together—thus *Prt* means $P \times r \times t$.

We have also seen (p. 30) that a convenient symbolism used is that, for instance, $a \times a \times a \times a$ is written a^4, the little 4 placed high and to the right of the *a* indicating that the product of a four times is to be taken. The reader should look over pp. 28–30 and 112, 113 before proceeding farther.

SUBSTITUTING IN A FORMULA

There are two well-known scales on which temperature is measured: the Fahrenheit scale (used in ordinary and clinical thermometers), and the Centigrade scale (generally used for scientific purposes).

It can be shown that if $F°$ and $C°$ represent the same temperature on the Fahrenheit and Centigrade scales respectively F and C are connected by the relation or formula:

$$F = \frac{9C}{5} + 32.$$

This formula is the algebraic way of writing down the fact that to obtain the temperature F on the Fahrenheit scale we multiply the corresponding temperature C on the Centigrade scale by nine, divide the result by five, and add 32 to the answer.

If we give C any numerical value this formula enables us to obtain the corresponding value of F. This process is called 'substituting' in a formula.

Thus, if $C = 10$, $F = \dfrac{9 \times 10}{5} + 32 = 18 + 32 = 50$.

Notice that when we change back to arithmetic—that is, replace letters by numbers—we re-insert the $\times$ sign.

And if $C = 100$, $F = \dfrac{9 \times 100}{5} + 32 = 180 + 32 = 212$.

And if $C = 0$, $F = 32$.

CHANGING A FORMULA

With scarcely any knowledge of algebra we can change our formula round so that it will enable us to find C when we know F. To do this we build up our F formula, starting from C and writing down in full what we are doing at each stage. This is done below, the steps being numbered for convenience; at each step the stage reached is given in square brackets. Adjacent to this, for convenience, are steps for the reverse process. Notice that the opposite of each command is given—'subtract' for 'add,' and so on.

1. Take C . . $[C]$ 1. Take F . . $[F]$

2. Multiply by 9 . $[9C]$ 2. Subtract 32 . $[F - 32]$

3. Divide by 5 . $\left[\dfrac{9C}{5}\right]$ 3. Multiply by 5 $[5(F - 32)]$
(Note use of brackets.)

4. Add 32 . $\left[\dfrac{9C}{5} + 32\right]$ 4. Divide by 9 $\left[\dfrac{5}{9}(F - 32)\right]$

This gives F.   This gives C.

So $C = \frac{5}{9}(F - 32)$.

Rather amusing, isn't it? Of course, the work is made very much shorter by the use of purely algebraic methods. We shall therefore next investigate, largely by diagrammatic or visual methods, the fundamental rules for algebraic operations. We shall then be prepared to go on to the solution of equations and their application to problems.

EXERCISE 26

Substitution

1. $A = 2h(l + b)$; $h = 12$, $l = 20$, $b = 16$. Find A.

2. $V = \pi r^2 h$; $r = 8$, $h = 70$, $\pi = \frac{22}{7}$. Find V.

3. $C = \frac{5}{9}(F - 32)$; $F = 122$. Find C.

4. $H = \dfrac{PLAN}{33,000}$; $P = 350$, $L = 22$, $A = 180$, $N = 50$. Find H.

5. $x = \dfrac{17t - 60}{4}$. Find x if $t = 4, 5, 6$ (three answers).

6. $h = ut - \frac{1}{2}gt^2$; $u = 2000$, $t = 10$, $g = 32$. Find h.

7. $v^2 = u^2 + 2gs$; $u = 16$, $g = 32$, $s = 5$. Find v.

8. $\theta = 15 - \cdot0065\ h$. Find θ if $h = 2000$.

9. $W = \dfrac{CAV^2}{196}$; $C = \cdot44$, $A = 1078$, $V = 100$. Find W.

Reversing of the Rule or Changing a Formula

10. $A = lb$. Find l.

11. $PV = RT$. Find V.

12. $H = \dfrac{PLAN}{33,000}$. Find P.

13. $A = \pi r^2$. Find r.

14. $s = \frac{1}{2}(u + v)t$. Find t.

15. $V = \pi ab(R - r)$. Find R.

16. $s = ut + \frac{1}{2}ft^2$. Find (i) u, (ii) f.

FUNDAMENTAL OPERATIONS

There are very few and simple rules which have to be learned and followed for working with symbols. To begin with, the following diagrams have been drawn to illustrate that these rules are, in fact, true. In them A, B, C, and a and b may be any size and width.

REMOVAL OF BRACKETS

Cut along the continuous (unbroken) lines and fit together:

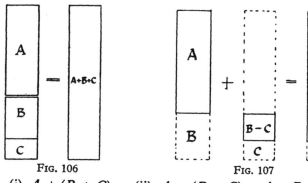

FIG. 106   FIG. 107

(i) $A + (B + C)$
$= A + B + C$.

(ii) $A + (B - C) = A + B - C$.

211

(iii) $A - (B + C) = A - B - C.$

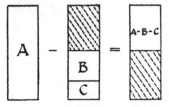

FIG. 108

(iv) $A - (B - C) = A - B + C.$

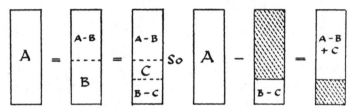

FIG. 109

The last example may be argued with numbers as follows:

To show that $15 - (7 - 3) = 15 - 7 + 3.$ $15 - 7$ is 7 less than 15. Now, if we subtract $(7 - 3)$ from 15 we are taking away 3 less than 7 from 15, and so the answer is 3 more than if we subtracted only 7 from 15.

In words the four rules become:

(i) A plus bracket B plus C bracket is equal to A plus B plus $C.$

(ii) A plus bracket B minus C bracket is equal to A plus B minus $C.$

(iii) A minus bracket B plus C bracket is equal to A minus B minus $C.$

(iv) A minus bracket B minus C bracket is equal to A minus B plus $C.$

Any letters or numbers may be used instead of A, B,

and C here. For example, if we write 3 for A, x for B, and y for C, we have:

$$3 + (x + y) = 3 + x + y.$$
$$3 + (x - y) = 3 + x - y.$$
$$3 - (x + y) = 3 - x - y.$$
$$3 - (x - y) = 3 - x + y.$$

MULTIPLICATION

Next we come to two simple multiplication rules. They are illustrated by the diagrams in Figs. 110 and 111.

Cut along the continuous (unbroken) lines and fit together.

(i) $a(a + b) = a^2 + ab.$

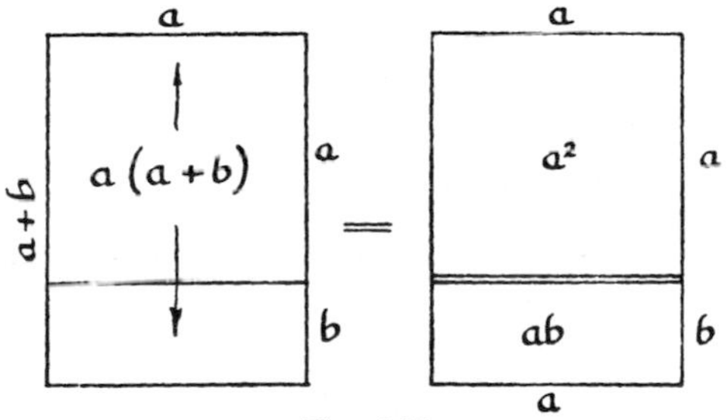

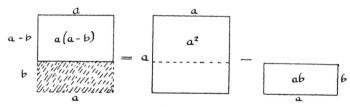

FIG. 110

(ii) $a(a - b) = a^2 - ab.$

FIG. 111

In words, these become:

(i) a bracket a plus b bracket is equal to a squared plus ab.

213

(ii) *a* bracket *a* minus *b* bracket is equal to *a* squared minus *ab*.

Note. Small letters have been used here for a change.

It is easy to see (draw diagrams for yourself) that we may similarly illustrate:

$$a(b + c) = ab + ac.$$
$$a(b - c) = ab - ac.$$

If we write, as an example, *x* for *a*, 5 for *b*, *y* for *c*, we have:

$$x(x + 5) = x^2 + 5x.$$
$$x(x - 5) = x^2 - 5x.$$
$$x(5 + y) = 5x + xy \text{ (notice we write } 5x,$$
$$x(5 - y) = 5x - xy. \qquad \text{not } x5).$$

Similarly, $2(5l + 7m) = 10l + 14m.$
$$16(3u - 10v) = 48u - 160v.$$

Remember that an expression like $5x + 4y$ cannot be simplified any further; if you add 5 apples to 4 pears you do not get 9 apples.

SQUARES

Finally, we have three rather more difficult operations.

(i) $(a + b)^2 = a^2 + 2ab + b^2$. (Fig. 112)

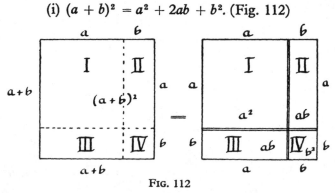

FIG. 112

214

Cut along the continuous (unbroken) line and fit together.

(ii) $(a - b)^2 = a^2 - 2ab + b^2$. (Fig. 113)

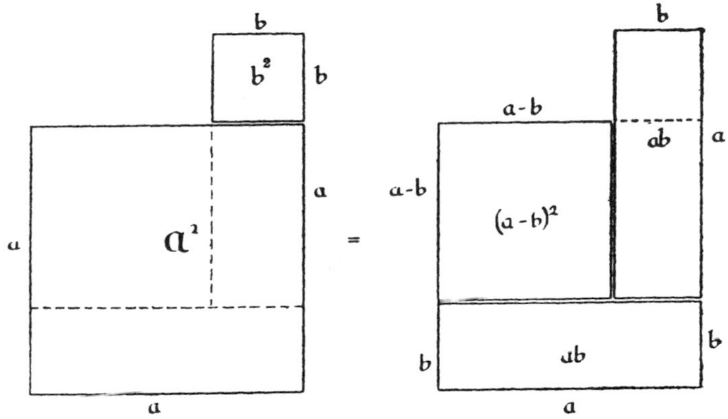

FIG. 113

The diagrams show that if ab and ab are taken away from $a^2 + b^2$ the remainder is $(a - b)^2$, so $a^2 + b^2 - 2ab = (a - b)^2$.

(iii) $a^2 - b^2 = (a + b)(a - b)$. (Fig. 114)

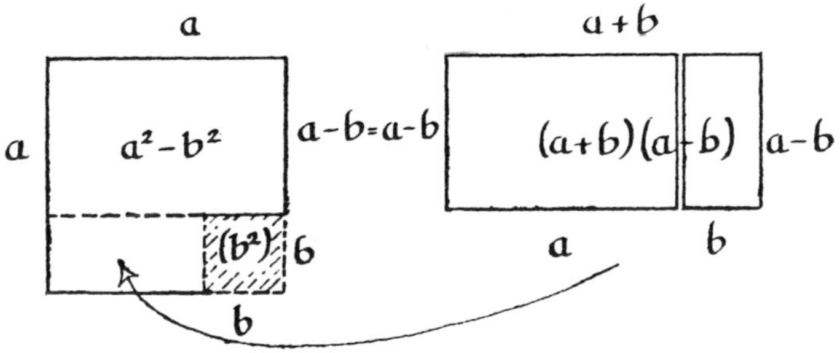

FIG. 114

The diagrams show that if b^2 is taken away from a^2, the remainder may be rearranged to form $(a + b)(a - b)$.

215

In words, these become:

(i) Bracket a plus b bracket all squared is equal to a squared plus $2ab$ plus b squared.

(ii) Bracket a minus b bracket all squared is equal to a squared minus $2ab$ plus b squared.

(iii) a squared minus b squared is equal to bracket a plus b bracket multiplied by bracket a minus b bracket.

The first two of these may be combined into one rule, very convenient for learning by heart, as follows:

First plus or minus last all squared is equal to first squared plus last squared plus or minus twice the product.

(Plus with plus, of course, and minus with minus.)

Learning these rules in this form makes one more independent of particular letters.

If we write x for a and 6 for b, we have:

$$(x + 6)^2 = x^2 + 12x + 36.$$
$$(x - 6)^2 = x^2 - 12x + 36.$$
$$x^2 - 36 = (x + 6)(x - 6).$$

Note also that $(x + 6)(x - 6) = x^2 - 36$.

Remember, once again, that in all these expressions any letters or numbers may be put for A, B, C, a, b, or c.

Practical Application. We can use any of these last three formulæ for simplifying the squaring of a number. Take, for example, the square of 57:

Using $\quad (a + b)^2 = a^2 + 2ab + b^2$,
and putting $a = 50$, $b = 7$, we have:
$$57^2 = 50^2 + (2 \times 50 \times 7) + 7^2$$
$$= 2500 + 700 + 49$$
$$= 3249.$$

Or, using $(a - b)^2 = a^2 - 2ab + b^2$,
    and putting $a = 60$, $b = 3$, we have:
$$57^2 = 60^2 - (2 \times 60 \times 3) + 3^2$$
$$= 3600 - 360 + 9$$
$$= 3609 - 360$$
$$= 3249.$$

Or, since from Fig. 114 we may also write:
$$a^2 = (a + b)(a - b) + b^2,$$
    and putting $a = 57$, $b = 3$ (to make $a + b = 60$),
we have: $$57^2 = (57 + 3)(57 - 3) + 3^2$$
$$= (60 \times 54) + 9$$
$$= 3240 + 9$$
$$= 3249.$$

As a variation we may also put $b = 7$, to make $a - b = 50$. The point about this simplification is that 'long' multiplication is dispensed with. Try it for several squares, using one method as a check on another.

EXERCISE 27

1. Simplify, by removing the brackets:

   $5 + (x + y)$; $17y + (a - 4)$; $30 + (16l - 3t)$; $3a + (4b - 6 + 5l)$; $2r + 3s + (8 - 5a - 7c)$; $4 - (a + b)$; $7 - (2x + 3y)$; $y - (x + 2)$; $2u - 3v - (15 - 2w)$; $3a - (4b + c) + (11e - f)$.

2. Multiply out:

   $4(x + 3)$; $7(y - 5)$; $3(2a + 9)$; $10(8 - 3t)$; $a(a + 17)$; $a(2a + 3)$; $b(5b - a)$; $3x(2x - 15)$; $7(11 + 3A) + 2B(5 - C)$; $14A(A + B) - 3(18 - B)$.

3. Write out in full:

   $(a + 1)^2$; $(b - 5)^2$; $(x + 2)^2$; $(2t - 3)^2$; $(5x + 4)^2$; $(x + 3y)^2$; $(3a - 5b)^2$; $(x + 2)(x - 2)$; $(2t + 3)(2t - 3)$; $(3 - 7x)(3 + 7x)$.

4. Draw diagrams to show that:

$$(a + b)(c + d) = a(c + d) + b(c + d)$$
$$= ac + ad + bc + bd.$$
$$(a + b)(c - d) = a(c - d) + b(c - d)$$
$$= ac - ad + bc - bd.$$
$$(a - b)(c - d) = a(c - d) - b(c - d)$$
$$= ac - ad - bc + bd.$$

5. Use the results of (4) to multiply out:

$(a + 3)(b + 2)$; $(x + 4)(y - 7)$; $(a - 2)(b - 5)$; $(3 + b)(2 + a)$; $(5 - x)(12 + t)$.

6. Use the methods of pp. 216, 217 to obtain, and check, the squares of: 49, 73, 85, 99, 103, 124.

CHAPTER XVI

EQUATIONS AND PROBLEMS

DO YOU LIKE SOLVING PUZZLES?

MOST of us are familiar with puzzle questions of the 'find the number' type. Their solution has provided men and women with recreation throughout the ages. Diophantus, whom we have previously mentioned, is said to have summed up his life in the problem:

> His boyhood lasted one-sixth of his life; his beard grew after one-twelfth more; he married after one-seventh more; and his son was born five years later. The son lived to half his father's age, and the father died four years after his son.

Such a problem would probably have been solved by a 'trial and error' method—or the 'rule of false,' as it was called. The man in the train and the solver of Sunday-paper puzzles would probably tackle it from the same angle; a person familiar with algebra, however, would condense the statement—translate it, if you like— by saying, "Suppose he died at the age of x years," and writing down the long statement algebraically as:

$$\tfrac{1}{6}x + \tfrac{1}{12}x + \tfrac{1}{7}x + 5 + \tfrac{1}{2}x + 4 = x.$$

Actually we find that Diophantus lived to be 84. (Check this.) Our business now is how to find this out from the algebraic expression—or *equation*, as it is called, since it expresses the fact that two quantities are equal to each other.

WHAT IS AN EQUATION?

We have already shown (p. 214) that:

$$x(x - 5) = x^2 - 5x.$$

This holds for all values of x (try it for yourself for a few different values of x, positive and negative); the right-hand side of the expression is in fact just a re-arranged form of the left-hand side.

Now consider the statement

$$x - 5 = 0.$$

If we put $x = 1$ in the expression $x - 5$ we have $x - 5 = 1 - 5$, or $- 4$; so it is not equal to 0. The same applies for $x = 2, 3,$ and 4. But if we put $x = 5$, then $x - 5 = 5 - 5 = 0$. If we carry on and try any other value for x, positive or negative, other than 5 we shall find that $x - 5$ is not equal to 0.

So the statement $x - 5 = 0$ is only valid, or true, when $x = 5$. This is a very elementary example of an equation. Equations in general are combinations of one or more letters and (or) numbers lying on each side of an $=$ sign; they are valid or true for (technically we say that they are satisfied by) certain values of the letters only. The letters are called the *unknowns*, or the *variables*. In this they differ from *identities* which, like

$$x(x - 5) = x^2 - 5x,$$

are combinations of a letter, or letters, and numbers which are such that the expression on one side of the $=$ is just a rearrangement of that on the other side. Identities are valid for *all* values of the letters.

Solving an equation or equations is the process of finding the value or values of the letter or letters which make the equation or equations true—in other words, finding the value or values of the letter or letters which satisfy the equation.

In particular, a *simple equation* is one in which the value of only *one* unknown letter (or *quantity*, as it is sometimes called) has to be found, and which is satisfied for only *one* value of that letter. In the course of our reading we shall meet other types of equations. But we shall start by considering the solving, or solution, of simple equations.

HOW TO SOLVE A SIMPLE EQUATION

The solution of simple equations presented considerable difficulty until the use of signs and symbols became universal. Those who have not forgotten all their algebra may care to reflect what a business it would be to have to write out in words all the operations entailed.

It is suggested that the reader who merely wishes to revise his knowledge of solving simple equations may care to omit the following illustrations of the rules and pass on directly to the summary at p. 224.

(i) If three things of equal weight, x lb., placed in one pan of a scales require 6 lb. in the other pan to balance them, then we have the equation:

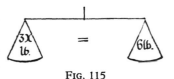

FIG. 115

$$3x = 6.$$

Since $3x$ balances 6 lb., then $\frac{1}{3}$ of $3x$ balances $\frac{1}{3}$ of 6 lb.—that is, x balances 2 lb., or $x = 2$.

We thus have:

Each side of an equation may be divided by the same number without upsetting the balance.

N.B. In arithmetic $3\frac{1}{6} = 3 + \frac{1}{6}$.
In algebra $3x = 3 \times x$.

(ii) (*a*) Suppose now that we have only an 8-lb. and a 2-lb. weight. Then we can balance the three things of equal weight (*x* lb.) against the 8-lb. weight by putting the 2-lb. weight in the pan with them.

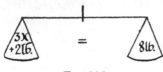

FIG. 116

This gives the equation $3x + 2 = 8$.

The balance would still be maintained if we could remove 2 lb. from each pan. This would give us an equation:

$$3x = 8 - 2 = 6,$$

and so, as before, $x = 2$.

(*b*) Suppose now that we have seven things of equal weight, 2 lb., and that we have only a 6-lb. weight. Then the balance will be secured if five of the things are put in the one pan, and the other two are put with the 6-lb. weight in the other pan.

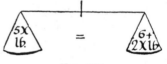

FIG. 117

This gives the equation $5x = 6 + 2x$.

The balance would still be maintained if we removed two of the things from each side. This would give us the equation:

$$5x - 2x = 6 + 2x - 2x = 6,$$

$$\text{or} \qquad 3x = 6,$$

$$\text{and} \qquad x = 2, \text{ by (i) above.}$$

In (ii) (*a*) and (*b*) we have illustrations of:

Any positive number or quantity may be removed to the other side of the equation if its sign is made negative.

222

(iii) Suppose now that we have again three of the things, but that a chip weighing 1 lb. has been cut off one of them; then 5 lb. in the other pan will be sufficient to balance them.

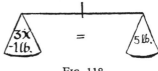

Fig. 118

This gives the equation $3x - 1 = 5$.

The balance would still be maintained if 1 lb. could be added to each side. This would give us the equation:

$$3x - 1 + 1 = 5 + 1,$$

or

$$3x = 5 + 1$$
$$= 6,$$

giving

$$x = 2, \text{ by (i) above.}$$

So our last rule may be extended thus:

Any number or quantity may be removed to the other side of the equation if its sign is changed.

(iv) There is one more rule which we require. We have already seen that if the two sides of an equation balance they will still balance if the two sides are each divided by the same number. It stands to reason also that if the two sides are multiplied by the same number they will still balance. For example:

if $\dfrac{x}{3} = 12$

$x = 36$ (multiplying each side by 3).

Or, if $\dfrac{3x}{4} = 9,$

$3x = 36$ (multiplying each side by 4),

so $x = 12$ (dividing each side by 3).

223

Here are the rules collected together:

We can multiply or divide both sides of an equation by the same number without upsetting the balance. We can add to or subtract any number or quantity from each side of an equation without upsetting the balance, or, what comes to the same thing, we can transfer any number or quantity from one side of the equation to the other if we change its sign (+ to −, −− to +) when we do so.

With the use of these rules the solving of an equation is done by bringing all the expressions containing the letter which we want to find to one side of the equation and taking everything else to the other side of the equation. Before doing this we may also have to use the other rules of algebra which we have learned in Chapter xv.

For example:

(a) Solve the equation $2(4n - 3) = 18$.
Divide through by 2. $\qquad\qquad 4n - 3 = 9$.
Transfer -3 to the other side
of the 'equals.' $\qquad\qquad\qquad 4n = 9 + 3 = 12$.
Divide through by 4. $\qquad\qquad\quad n = 3$.
$\qquad$*Check.* $\qquad\qquad 2(4n - 3) = 2(12 - 3)$
$\qquad\qquad\qquad\qquad\qquad\qquad = 2(9)$
$\qquad\qquad\qquad\qquad\qquad\qquad = 18$.

Note. Always check your answer by substituting the value which you have found and showing that it satisfies the equation—that is, makes the two sides of the equation equal to one another.

(b) In a certain machine if an effort E is required to raise a load L, then $E = \frac{1}{5}L + 2\frac{1}{4}$. Find the load which will be raised by an effort of 40 lb.

We have to solve the equation $\qquad 40 = \frac{1}{5}L + 2\frac{1}{4}$.

Interchange the sides to avoid
bringing $\frac{1}{5}L$ across the 'equals.' $\frac{1}{5}L + 2\frac{1}{4} = 40$.

Take the $2\frac{1}{4}$ to the other side of the equals.

$$\tfrac{1}{5}L = 40 - 2\tfrac{1}{4}$$
$$= 37\tfrac{3}{4}.$$

Multiply each side by 5. $\quad L = 188\tfrac{3}{4}$ lb.

Check. $\qquad E = 40.$

$$\tfrac{1}{5}L + 2\tfrac{1}{4} = \tfrac{1}{5}(188\tfrac{3}{4}) + 2\tfrac{1}{4}$$
$$= 37\tfrac{3}{4} + 2\tfrac{1}{4}$$
$$= 40.$$

(*c*) In a problem on weight distribution after substituting in a formula the resulting equation was:

$$W - \frac{W}{2} = 20 - \frac{W}{3}.$$

Find the value of W.

We multiply each side by the L.C.M. of 2 and 3 (that is, 6) in order to get rid of the fractions. We are justified in doing this, for as the sides of the equation are equal to each other they will still remain equal if we multiply each of them by 6.

This gives $\quad 6W - 3W = 120 - 2W.$

Bring all the terms containing W to the left-hand side.

$$6W - 3W + 2W = 120.$$

So $\qquad\qquad 5W = 120.$

$$W = 24.$$

Check. $\qquad W - \dfrac{W}{2} = 24 - 12$
$$= 12.$$
$$20 - \frac{W}{3} = 20 - 8$$
$$= 12.$$

(*d*) In determining the value of a force P, the following equation was obtained.

$$\frac{3(P + 2)}{4} - \frac{2(4 + P)}{5} = 2.$$

Find P.

P

Multiply through by the L.C.M. of 4 and 5 (that is, 20) to get rid of fractions.

$$15(P + 2) - 8(4 + P) = 40.$$
$$\therefore \ 15P + 30 - 32 - 8P = 40. \quad \text{(On multiplying out}$$

(On multiplying out a bracket the minus sign in front changes the sign of each term in it.)

$$15P - 8P = 40 - 30 + 32.$$
$$7P = 72 - 30 = 42.$$
$$P = 6.$$

Check. $\dfrac{3(P + 2)}{4} - \dfrac{2(4 + P)}{5} = \dfrac{3(6 + 2)}{4} - \dfrac{2(4 + 6)}{5}$

$$= \dfrac{3(8)}{4} - \dfrac{2(10)}{5}$$
$$= 6 - 4$$
$$= 2.$$

Note.

If $\quad 5t = -30$
$\quad\quad t = -6.$
If $\ -3a = \quad 36$
$\quad\quad a = -12.$
If $\ -8n = -44$
$\quad\quad n = \quad 5\frac{1}{2}.$

EXERCISE 28

After substituting in various formulæ the following equations were obtained. Solve them to find the value of the unknown letter in each case.

1. $2n - 3 = 17.$
2. $6x + 12 = 4x + 7.$
3. $7P + 20 = 5P - 14.$
4. $3(F - 7) = 18.$
5. $6(2L - 3) = 5(4L - 6).$
6. $2(3v - 10) = 5(v + 9) - 68.$
7. $4(7y + 6) - 3(20y + 17) = 5.$
8. $\dfrac{4x}{3} = 18.$
9. $\dfrac{5z - 17}{3} = 21.$

10. $\dfrac{2x}{3} = \dfrac{3x}{2} + \dfrac{15}{4}$.

11. $\dfrac{3n}{4} - 2 = \dfrac{5n}{2} + 2\tfrac{2}{3}$. (*Hint.* $2\tfrac{2}{3} = \tfrac{8}{3}$.)

12. $\dfrac{P-1}{6} + \dfrac{2P+1}{5} = 4$.

13. $\dfrac{2(2r+3)}{7} - \dfrac{3(5r-2)}{4} + 4 = 0$.

14. $\cdot 75t = 3$. (*Hint.* Convert decimals into fractions.)

15. $2\cdot37T + 11\cdot6 = 1\cdot25T + 34$.

16. Using $\dfrac{9C}{5} = F - 32$, find C when $F = 50$.

17. From $Ftg = W(v - u)$ find v when $F = 18$, $t = 5$, $g = 32$, $W = 80$, $u = 63$.

18. From $s = ut + \tfrac{1}{2}gt^2$ find u when $s = 6000$, $g = 32$, $t = 10$.

19. From $n(R - r) = 2R$ find R when $n = 8$, $r = 4$.

20. Using $V - v = -e(U - u)$ find u when $V = 29$, $v = 11$, $U = 34$, $e = \cdot6$

21. Using $H = \dfrac{2\pi r n}{33000}$ find n when $H = 800$, $\pi = 3\tfrac{1}{7}$, $r = 3\cdot5$.

22. Using $s = \tfrac{1}{2}(u + v)t$ find u when $s = 5000$, $t = 25$, $v = 88$.

23. A shell whose speed is u ft. per second will penetrate a distance of d in. into a plate where $d = \dfrac{15u}{6000 - u}$. Find u, to nearest whole number, if $d = 6$ in.

PROBLEMS LEADING TO SIMPLE EQUATIONS

Simple equations are often met in the solution of problems as well as after substitution in formulæ. Consider the following examples:

(*a*) A 'puzzle' question. Find a number which, when increased by 4 and the result multiplied by 7, is equal to the amount by which twice the original number is less than 100.

Let n be the number.

Increase it by 4.	$n + 4$.
Multiply the result by 7.	$7(n + 4)$.
Start with 100.	100.
Subtract twice the original number.	$100 - 2n$.

The statement in the example gives the equation:

$$7(n + 4) = 100 - 2n.$$

Solving this, $\quad 7n + 28 = 100 - 2n.$

$$7n + 2n = 100 - 28.$$
$$9n = 72.$$
$$n = 8.$$

Check. $7 \text{ (number} + 4) = 7(8 + 4) = 7(12) = 84.$
$100 - 2 \text{ (number)} = 100 - 16 = 84.$

(*b*) The cruising speed of an aeroplane is 300 m.p.h.; its maximum speed is 360 m.p.h. A flight of 320 miles has to be completed in an hour. How far can the airman go at cruising speed?

Suppose he flies a distance of d miles at cruising speed. Then he flies a distance of $320 - d$ miles at top speed.

Time taken at cruising speed $= \dfrac{d}{300}$ hours.

Time taken at top speed $\quad = \dfrac{320 - d}{360}$ hours.

But the total time $\quad\quad\quad = 1$ hour.

So $\quad\quad \dfrac{d}{300} + \dfrac{320 - d}{360} = 1.$

The L.C.M. of 300 and 360 is 1800. Multiply through by 1800.

$$6d + 5(320 - d) = 1800.$$
$$6d + 1600 - 5d = 1800.$$
$$d = 1800 - 1600$$
$$= 200 \text{ miles.}$$

Check. $\qquad$ Cruising time $= \frac{200}{300} = \frac{2}{3}$ hour.

Time at top speed $= \frac{120}{360} = \frac{1}{3}$ hour.

Total time $= \frac{2}{3} + \frac{1}{3} = 1$ hour.

(*c*) A silver collection was made at the end of a show. The total proceeds were £9 4*s.*, and there were 208 coins in all, all of them sixpences or shillings. How many of each coin were there?

Suppose there were *x* shillings.

Then there were 208 − *x* sixpences.

The value of the sixpences is $\frac{1}{2}(208 - x)$ shillings.

So the total value of the coins is

$$x + \tfrac{1}{2}(208 - x) \text{ shillings.}$$

But this amounts to 184 shillings.

The corresponding equation is therefore:

$$x + \tfrac{1}{2}(208 - x) = 184.$$

Multiply through by 2 to clear of fractions:

$$2x + 208 - x = 368.$$
$$x = 368 - 208 = 160.$$
$$\text{So} \quad 208 - x = 208 - 160 = 48.$$

Check. $\quad$ 160 shillings $\qquad\qquad = $ £8.

48 sixpences $= 24$ shillings $= $ £1 4*s.*

Total $= $ £9 4*s.*

Note. The answers to problems should always be checked from the words of the problem, not from the equations.

EXERCISE 29

1. Take a number; multiply it by 5; subtract 18 from the answer, and the result is 17. What is the number?

2. Take a number; add 8 to it; double the answer, and the result is 54. What is the number?

3. Take 27 from three times the number I am thinking of, and you have three-quarters of my number. What is my number?

4. How much must be added to 30 so that the added part will be one-quarter of the resulting number?

5. 420 cabbage-plants are to be planted out in 18 rows. Some rows will have 25 plants, the rest 20 plants. How many rows of 25 are there?

6. A man is four times as old as his daughter. In six years' time he will only be three times as old as his daughter. How old is each of them now?

7. One train travelling between two places at 48 m.p.h. does the journey in 40 minutes less than a train which averages 24 m.p.h. How far apart are the two places? (Hint: work in hours.)

8. 240 people saw a show, some in half-crown seats and the rest in 1s. 9d. seats. The takings were £24 15s. How many paid half a crown?

9. A gardener reckons that for every seed potato which he sets he will later dig up seven. If he counts 7 potatoes to the lb. in the crop how many potatoes must he plant if he wants 3 cwt. for eating and enough left over to set the same number next year?

10. A motorist averages 20 m.p.h. when passing through a built-up area and 30 m.p.h. in the open country. A journey of 70 miles took him 2½ hr. For how many miles of it was he passing through built-up areas?

11. A man normally earns £1 a day, but some days he puts in longer time and gets 24s. After 12 working days he had earned £13 8s. On how many days did he work longer?

12. Divide 15s. between three boys, Tom, Dick, and Harry, so that Tom gets twice as much as Dick and Harry gets a shilling more than twice what Tom gets.

CHAPTER XVII

MORE EQUATIONS AND PROBLEMS

SIMULTANEOUS EQUATIONS

SUPPOSE we are given the following problem to solve:

100 people attended a dance; the men paid 3s. each, the women 2s. If the entrance money came to £12, how many men and how many women were present?

Here we have to find two unknown things—the number of men and the number of women. But we are given two facts: the total number of people present and the total money paid by them. From these two facts we can form two equations as follows.

Suppose the number of men is x and the number of women is y. Then, since there are 100 people present, we must have

$$x + y = 100.$$

This is our first equation.

Now x men pay $3x$ shillings altogether, and y women pay $2y$ shillings altogether. So, since £12 = 240 shillings, and both sides of an equation must be expressed in the same units:

$$3x + 2y = 240.$$

This is our second equation.

We thus have two equations to be solved together. (When there are two unknown letters we require two equations for finding them.) They cannot be solved separately, for we cannot find the value of x unless we know the value of y, and *vice versa*. Such equations are called *simultaneous equations*, for they are both true for the same values of the letters, and they can only be solved simultaneously—that is, at the same time, or when taken together. We bracket them together thus:

$$\left.\begin{array}{r} x + y = 100 \\ 3x + 2y = 240 \end{array}\right\}.$$

To solve the equations we multiply either or both by such numbers as will make the *coefficients* of (that is, the numbers in front of) x or y equal. In this particular case if we multiply both sides of the first equation by 2 and leave the other equation as it is we make the coefficients of y equal. We then have:

$$\left.\begin{array}{r} 2x + 2y = 200 \\ 3x + 2y = 240 \end{array}\right\}.$$

Subtracting, $\qquad -x \qquad = -40,$

and so $\qquad x = 40.$

To find y take either equation, in the original unmultiplied form, and substitute this value of x in it.

Taking $\qquad x + y = 100,$

we have $\quad 40 + y = 100.$

So $\qquad\qquad y = 100 - 40 = 60.$

We check our answer in the *other* equation.

Substituting $x = 40$, $y = 60$ in the left-hand side of $3x + 2y = 240$, we have

$$3x + 2y = 120 + 120$$
$$= 240;$$

and so $x = 40$, $y = 60$ satisfies both equations.

Note. In *problems* giving rise to simultaneous equations we also check our answer in the problem itself. Thus here the total number of people $= 40 + 60 = 100$. The men pay $40 \times 3s. = 120s. = £6$; the women pay $60 \times 2s. = 120s. = £6$; so the takings are £12. Of course we are not asked to solve the problem of how 40 men can dance with 60 women!

We have in algebra problems another illustration of the principle of abstraction in mathematics. As we have said before, the good mathematician is the good abstractor; and in the writing down of an equation, or equations, for the solution of problems we have abstraction at a high level.

We now give some further examples of solving pairs of simultaneous equations.

(a) *Solve*
$$\left.\begin{array}{r} 5u + 6v = 28 \\ 7u - 4v = 2 \end{array}\right\}.$$

Here we can make the coefficients of u equal by multiplying both sides of the first equation by 7 and of the second by 5. Thus:

$$\left.\begin{array}{r} 35u + 42v = 196 \\ 35u - 20v = 10 \end{array}\right\}$$

Subtracting, $\qquad\qquad 62v = 186.$

So $\qquad\qquad\qquad\quad v = 3.$

Substituting $v = 3$ in $5u + 6v = 28,$

$$5u + 18 = 28.$$
$$5u = 10.$$
$$u = 2.$$

Checking in the second equation,

$$7u - 4v = 14 - 12 = 2.$$

(b) *Solve*
$$\left.\begin{array}{l} \dfrac{m}{3} - \dfrac{n}{4} = 6 \\[2mm] \dfrac{m}{4} + \dfrac{n}{10} = 2\tfrac{1}{5} \end{array}\right\}.$$

First clear of fractions by multiplying both sides of each equation by the L.C.M. of the denominators of its terms. This gives:

$$\left.\begin{array}{r} 4m - 3n = 72 \\ 5m + 2n = 44 \end{array}\right\}.$$

233

Here we shall make the coefficients of n equal by multiplying both sides of the equations by 2 and 3 respectively, giving:

$$8m - 6n = 144$$
$$15m + 6n = 132$$

We get rid of n by *adding* (since $- 6n + 6n = 0$).

Then $23m = 276.$
                $m = 12.$

Substituting this value of m in the second equation (by way of change, either equation will do):

$$60 + 2n = 44.$$
$$2n = 44 - 60 = - 16.$$
$$n = - 8.$$

Check in the original first equation.

$$\frac{m}{3} - \frac{n}{4} = 4 - (- 2)$$
$$= 4 + 2 = 6.$$

Note. If the coefficients contain decimal fractions make them whole numbers first by multiplying through by 10, 100, or such power of 10 as is necessary.

A GENERALIZATION

In this section we have considered only simultaneous equations in two variables. In certain kinds of work equations in more variables than two have to be solved. The methods employed vary somewhat, but the following general principle always holds: if there are n variables n equations are necessary for their determination. Thus three equations are necessary if there are three unknowns, four for four unknowns, and so on.

EXERCISE 30

Solve the following pairs of simultaneous equations:

1. $\left.\begin{array}{l} x + y = 16 \\ x - y = 2 \end{array}\right\}$

2. $\left.\begin{array}{l} 2x + y = 8 \\ 3x + 5y = 19 \end{array}\right\}$

3. $\left.\begin{array}{l} 3P + 5Q = 370 \\ 5P + 3Q = 590 \end{array}\right\}$

4. $\left.\begin{array}{l} 3u - 2v = 19 \\ 2u + 3v = 43 \end{array}\right\}$

5. $\left.\begin{array}{l} 2r_1 - 3r_2 = 10 \\ 8r_1 + 7r_2 = 2 \end{array}\right\}$

6. $\left.\begin{array}{l} 3m - n = 30 \\ 5m + 7n = 310 \end{array}\right\}$

7. $\left.\begin{array}{l} 2a + 3b = 61 \\ 5a - 4b = 26 \end{array}\right\}$

8. $\left.\begin{array}{l} 4y - 5x = 58 \\ 7y - 3x = 67 \end{array}\right\}$

9. $\left.\begin{array}{l} 79\frac{1}{4} = 5r + 6s \\ 14 = 4r - 3s \end{array}\right\}$

10. $\left.\begin{array}{l} 3F_1 - 2F_2 = 0 \\ F_1 + F_2 = 11\frac{1}{4} \end{array}\right\}$

11. $\left.\begin{array}{l} 4.6x + 3y = 64 \\ 2.7x + 2y = 39 \end{array}\right\}$

12. $\left.\begin{array}{l} 3.2p - 7.5q = 7.4 \\ 11p + 5q = 87 \end{array}\right\}$

13. $\left.\begin{array}{l} 4u + 9v = 51 \\ 8u = 9 + 13v \end{array}\right\}.$

14. $\left.\begin{array}{l} 4c_1 + 5c_2 = 22 \\ \dfrac{c_1}{3} = \dfrac{c_2}{2} \end{array}\right\}$

15. $\left.\begin{array}{l} \dfrac{x}{2} + \dfrac{y}{3} = 29 \\[2mm] \dfrac{x}{3} - \dfrac{y}{2} = 2 \end{array}\right\}$

16. $\left.\begin{array}{l} \dfrac{l}{2} + \dfrac{m}{3} = 13 \\[2mm] \dfrac{l}{5} + \dfrac{m}{8} = 5 \end{array}\right\}$

PROBLEMS LEADING TO SIMULTANEOUS EQUATIONS

We have already solved one such problem at pp. 231, 232. Here are two more examples.

(a) A man set the following conundrum: three times my age is 51 years more than twice my wife's age; four times my age is 54 years more than three times my wife's age. What are our ages?

Suppose that the man is x years old and that his wife is y years old. Then, translating the facts into algebra, we have:

$$\left.\begin{array}{l} 3x - 2y = 51 \\ 4x - 3y = 54 \end{array}\right\}.$$

To solve these simultaneous equations we shall get rid

of y by multiplying the first line by 3, the second line by 2, and subtracting:

$$9x - 6y = 153$$
$$8x - 6y = 108$$
$$x \qquad = 45.$$

Substituting $x = 45$ in $3x - 2y = 51$, we have:
$$135 - 2y = 51.$$
$$- 2y = 51 - 135$$
$$= - 84.$$
$$y = 42.$$

So their ages are 45 and 42 years.

Check.

Three times man's age	$= 3 \times 45$	$= 135.$
Twice woman's age	$= 2 \times 42$	$= 84.$
Difference		$= 51.$
Four times man's age	$= 4 \times 45$	$= 180.$
Three times woman's age	$= 3 \times 42$	$= 126.$
Difference		$= 54.$

(b) In a simple machine if an effort of E lb. weight moves a load of L lb. weight, E and L are connected by a relation $E = aL + b$ (called the Law of the Machine) where a and b are constants for the machine. If efforts of 8 and 10 lb. weight move loads of 20 and 40 lb. weight respectively find the values of a and b. Using these values of a and b, find the load which will be moved by an effort of 11 lb. weight.

Substituting the given values of E and L we have:

$$8 = a \cdot 20 + b$$
and $$10 = a \cdot 40 + b$$ or $$20a + b = 8$$
$$40a + b = 10$$

$$\text{Subtracting,} \quad - 20a \quad = - 2.$$
$$\text{So} \qquad a = \tfrac{1}{10}.$$

Substituting in $20a + b = 8$,
$$\text{we have} \qquad 2 + b = 8.$$
$$b = 6.$$

Check. $\quad 40a + b = 4 + 6 = 10.$

So $\qquad E = \tfrac{1}{10} L + 6.$

Substituting $E = 11,$

we have $\quad 11 = \tfrac{1}{10} L + 6.$

$\qquad\qquad 5 = \tfrac{1}{10} L.$

$\qquad\qquad L = 50$ lb. weight.

EXERCISE 31

(Many of these problems can be solved by the use of only one letter. The first few are given for practice in forming easy equations.)

1. The sum of two numbers is 87; their difference is 15. Find the numbers. (Take x, y as the numbers.)

2. Four times one number is equal to five times another number. If twice the greater number added to three times the smaller is equal to 220 what are the numbers?

3. The sum of two numbers is five times their difference, and three times the greater number is 30 more than twice the smaller. Find the numbers.

4. Find two numbers which are such that one-third of the greater exceeds a quarter of the smaller by 8, and a half of the smaller is 8 more than one-sixth of the larger.

5. Two numbers are in the ratio 3 : 2. If 20 is added to each of them their ratio becomes 5 : 4. What are the numbers?

6. (A very old catch.) A bottle and its cork cost $2\tfrac{1}{2}d.$, and the bottle costs $2d.$ more than the cork. Find the cost of the cork.

7. A number of five-gallon drums of oil, some weighing 64 lb. and others weighing 80 lb. each, are mixed. There are 30 drums, and their total weight is 2112 lb. How many drums of each kind of oil are there?

8. How many gallons of lubricating oil weighing 10·4 lb. per gallon must be mixed with a heavier oil weighing 12·2 lb. per gallon to form 60 gallons of a mixture weighing 11·6 lb. per gallon?

9. The average weight of 50 men is 150 lb. The average weight of those who are less than 11 stone is 142 lb.; the average weight of those who are more than 11 stone is 167 lb. How many men weigh less than, and how many more than, 11 stone?

10. At a factory the ratio of the number of men to the number of women employed was 7 : 3. After five more men and one more woman were taken on the ratio became 12 : 5. How many men and women were there originally?

11. In a certain machine the effort (E lb.) and the load (L lb.) are connected by the relation $E = aL + b$, where a and b do not vary. When there is no load an effort of 60 lb. is necessary to put the parts of the machine in motion; and an effort of 100 lb. will raise a load of 800 lb.

   (i) What are the values of a and b?
   (ii) What effort raises a load equal to itself? (Nearest lb.)
   (iii) What effort raises a load ten times as great as itself?
   (iv) Will an effort of one cwt. be sufficient to raise a load of half a ton?

12. For a certain spring under compression the compressed length (l in.) and the pressure (P lb. per sq. in.) are connected by the formula $l = L - kP$, where L and k are always the same for this spring provided P does not exceed 40 lb.

   (i) Find L and k if $l = 10$ when $P = 20$ and $l = 12$ when $P = 12$.
   (ii) What is the length of the spring when it is not under compression?
   (iii) What *compression* is produced by a pressure of 30 lb. per sq. in.?
   (iv) What pressure produces a *compression* of one in.?

QUADRATIC EQUATIONS

Consider the equation

$$x^2 + x - 42 = 0.$$

By actual multiplication,

$$(x - 6)(x + 7) = x^2 - 6x + 7x - 42 = x^2 + x - 42.$$

$$\text{So if} \quad x^2 + x - 42 = 0,$$
$$\text{then} \quad (x - 6)(x + 7) = 0;$$

and this is true whether

$$x - 6 = 0 \text{ or } x + 7 = 0;$$
$$\text{that is, if} \quad x = 6 \text{ or} \quad x = -7.$$

238

Such an equation as this, in which there is only one unknown letter, the highest power of that letter being its square, is called a *quadratic equation*. It can be shown that a quadratic equation cannot be satisfied by more than two values of the unknown letter.

The method shown above can only be used for *certain* quadratic equations, but all that can be solved at all can be solved by means of the formula given below. It is not a difficult matter to show how the formula is obtained, but it is sufficient to know how to apply it to the solution of a quadratic equation. This is the formula:

If the equation is $ax^2 + bx + c = 0$

$$\text{then} \quad x = \frac{-b \pm \sqrt{b^2 - 4ac}}{2a}.$$

(For remembering, read it as: if $ax^2 + bx + c$ equals 0 then x equals $-b$ plus or minus the square root of $b^2 - 4ac$, all over $2a$.)

The $\pm$ means that:

$$x = \frac{-b + \sqrt{b^2 - 4ac}}{2a} \text{ or } x = \frac{-b - \sqrt{b^2 - 4ac}}{2a}.$$

We have to use the letters a, b, and c instead of numbers in the formula so as to make it applicable to *all* quadratic equations. These letters may represent positive or negative numbers, though usually the equation is written so that a is positive.

We shall now apply it to particular equations.

(a) *Solve* $x^2 + x - 42 = 0$ (the equation with which we commenced this section).

Compare it with $ax^2 + bx + c = 0$.

We see that $a = +1$, $b = +1$, $c = -42$.

So $\quad x = \dfrac{-b \pm \sqrt{b^2 - 4ac}}{2a}$ gives:

$$x = \frac{-1 \pm \sqrt{1 - 4(1)(-42)}}{2(1)}$$

$$= \frac{-1 \pm \sqrt{1 + 168}}{2} = \frac{-1 \pm \sqrt{169}}{2}$$

$$= \frac{-1 \pm 13}{2}$$

$$= \frac{-1 + 13}{2} \text{ or } \frac{-1 - 13}{2}$$

$$= \frac{12}{2} \text{ or } \frac{-14}{2} = 6 \text{ or } -7$$

Check.

For $x = 6$, $x^2 + x - 42 = 36 + 6 - 42 = 42 - 42 = 0$.
For $x = -7$, $x^2 + x - 42 = 49 - 7 - 42 = 49 - 49 = 0$.

(*b*) *Solve* $\quad 6t^2 - 19t + 15 = 0$.

Comparing it with $at^2 + bt + c = 0$ (note the change from x to t), we have

$$a = +6, b = -19, c = +15.$$

So $\quad t = \dfrac{-(-19) \pm \sqrt{(-19)^2 - 4(6)(15)}}{2(6)}$

$$= \frac{+19 \pm \sqrt{361 - 360}}{12}$$

$$= \frac{19 \pm \sqrt{1}}{12} = \frac{19 \pm 1}{12}$$

$$= \frac{19 + 1}{12} \text{ or } \frac{19 - 1}{12}$$

$$= \frac{20}{12} \text{ or } \frac{18}{12}$$

$$= \tfrac{5}{3} \text{ or } \tfrac{3}{2} = 1\tfrac{2}{3} \text{ or } 1\tfrac{1}{2}.$$

Check. For $t = \frac{5}{3}$, $6t^2 - 19t + 15$
$$= 6(\tfrac{25}{9}) - 19(\tfrac{5}{3}) + 15$$
$$= \frac{50}{3} - \frac{95}{3} + 15 = \frac{-45}{3} + 15 = 0.$$

For $t = \frac{3}{2}$, $6t^2 - 19t + 15$
$$= 6(\tfrac{9}{4}) - 19(\tfrac{3}{2}) + 15$$
$$= \tfrac{27}{2} - \tfrac{57}{2} + 15 = -\tfrac{30}{2} + 15 = 0.$$

We shall not solve quadratic equations in which the number under the root sign is not a perfect square. Later, when a method of extracting square roots has been learned, it will be realized that the general method of solving the equations is precisely the same. Of course, the checking is more laborious.

EXERCISE 32

Solve:

1. $x^2 - 3x + 2 = 0.$ 2. $x^2 + 7x + 10 = 0.$
3. $y^2 - 6y + 8 = 0.$ 4. $n^2 + 2n - 8 = 0.$
5. $2x^2 - 5x + 2 = 0.$ 6. $6t^2 + t - 12 = 0.$
7. $6u^2 + 11u - 10 = 0.$ 8. $2n^2 - 15n + 7 = 0.$

9. $2x^2 = 29x + 48.$ (First bring all the terms to the left-hand side of the 'equals.')

10. $n(n + 1) = 56.$ (First multiply out the brackets.)

PROBLEMS LEADING TO QUADRATIC EQUATIONS

As for the other problems with which we have dealt, the statement or statements given must first be translated into an equation in algebra. The equation is then solved, and the answers checked, as above. For example:

A party chartered a motor-coach for a trip at an agreed price of £7 10s. Five of them cried off, and so the others had to pay 1s. more than they expected to have to pay. What was the original number of people in the party?

Q

We must work in the same units throughout. Shillings will be most convenient here; £7 10s. = 150s.

Suppose x people meant to make the trip.

Then the share of each was $\dfrac{150s.}{x}$

But only $x - 5$ people actually went.

Their share was therefore $\dfrac{150s.}{x - 5}$

Expressing by algebra that this means 1s. more per person, we have:

$$\frac{150}{x - 5} = \frac{150}{x} + 1.$$

We must first clear this equation of fractions. To do this it will be sufficient to multiply both sides by the product of the denominators—that is, $(x - 5)x$.

This gives:

$$150x = 150(x - 5) + (x - 5)x;$$

that is, $\quad 150x = 150x - 750 + x^2 - 5x,$

or $-x^2 + 5x + 750 = 0$ (bringing all the terms to the left-hand side of the 'equals'),

or $\quad x^2 - 5x - 750 = 0$ (multiplying through by -1 to make the sign of x^2 positive).

$$x = \frac{5 \pm \sqrt{25 + 3000}}{2} = \frac{5 \pm \sqrt{3025}}{2}$$

$$= \frac{5 \pm 55}{2} \quad \begin{array}{l} (3025 = 25 \times 121; \\ \sqrt{3025} = 5 \times 11 = 55) \end{array}$$

$$= \frac{60}{2} \text{ or } -\frac{50}{2} = 30 \text{ or } -25.$$

So 30 people originally intended to go.

Check. If 30 people went, cost per head $= \dfrac{150}{30} = 5s.$

If 25 people went, cost per head $= \dfrac{150}{25} = 6s.$

Note. Clearly -25 people could not go. Yet -25 is a perfectly good solution *of the equation*; the reader should check this for himself.

EXERCISE 33

(All square roots work out exactly except in the last problem.)

1. The sum of a number and its square is 42. Find the number.

2. Two numbers differ by 8, and their product is 153. Find them. (Hint: take x and $x + 8$ as the numbers.)

3. Find two numbers whose sum is 22 and whose product is 120. (Hint: take x and $22 - x$ as the numbers.)

4. The sum of the first n whole numbers is $\frac{1}{2}n(n + 1)$.
   (i) Find the sum of all the whole numbers from 1 to 20.
   (ii) Find the sum of all the whole numbers between 41 and 100 inclusive. (Hint: first find the sum of the first forty.)
   (iii) How many whole numbers in the 'run' starting from 1 must be taken so that their sum is 105?

5. During an evening each of a party gave each of the others a cigarette. In all 132 cigarettes were given. How many were there in the party?

6. A party engaged a car for a journey for £3. Two did not turn up, and so the share of each of the others was 5s. more. How many had originally intended to go?

7. It is required to make a new panel which doubles the area of a panel 8 ft. by 6 ft. by adding a strip of the same width along two sides which lie next to each other. How wide must the strip be? Take the square root to the nearest whole number. (Hint: if the width of the new strip is x ft. what are the new length and breadth?)

CHAPTER XVIII

GRAPHS (CONTINUED): FUNCTIONS

A NEW KIND OF GRAPH

IN Chapter VIII we drew and interpreted graphs illustrating statistics and graphs useful for shortening calculations and saving time. We shall now show that a graph may also be used to illustrate the relationship between two quantities which are connected by an equation or formula.

We saw at p. 210 that the formula for converting temperatures ($F°$) on the Fahrenheit scale into temperatures ($C°$) on the Centigrade scale was:

$$C = \frac{5}{9}(F - 32).$$

Let us calculate and make a table of the values of C corresponding to some values of F. This is done by substituting the various values of F in the formula, and is best set out as follows:

F	−13	− 4	5	14	32	50	68	86	...	212
$F - 32$	−45	−36	−27	−18	0	18	36	54	...	180
$C = \frac{5}{9}(F - 32)$	−25	−20	−15	−10	0	10	20	30	...	100

The third row is formed by multiplying each term in the second row by $\frac{5}{9}$. The numbering of a scale must start somewhere, so negative temperatures are just regarded as temperatures below—that is, less than—zero, as we have previously seen when dealing with signed numbers. We have the same idea in numbering years. A.D. 1948 was the 1948th year after the birth of Christ; 55 B.C. was the 55th year before the birth of Christ, and could be written (though less conveniently) as − 55 A.D.

In drawing statistical graphs we dealt only with positive quantities, so we shall now explain how points to which negative numbers are affixed are plotted. Our reference above to 'below' and 'before' gives us the clue.

Draw two straight lines AOB, LOM at right angles to each other. Then, if Centigrade temperatures are measured along LOM, positive temperatures being measured along and in the direction of OL, negative Centigrade temperatures will be measured along and in the direction of OM. Similarly, positive and negative Fahrenheit temperatures are measured along and in the direction of OB and OA respectively. The signs of F and C in the four quadrants into which the axes AOB, LOM divide the plane of the paper are shown in the diagram (Fig. 119).

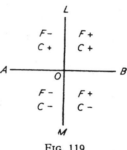

Fig. 119

For the points which we shall now plot the axis of F may be labelled from about $-15°$ to $+225°$ and the axis of C from about $-25°$ to $+100°$. This helps us to decide where to place the axes on the squared paper.

Plot the points, marking each by a small $\times$.

Join each point to the next point by straight lines; it will be found that they all lie on one straight line. It can be shown, by actual checking as well as by rigorous mathematical argument (which need not concern us here), that *any* point on this straight line (not only the plotted points) is such that its *co-ordinates*—that is, the number-pair formed by the values of F and C at that point—satisfy the given relation between F and C.

This is not an accident. When the relation connecting two variables is of the first degree—that is, contains

245

neither higher powers than the first of each variable nor the product of the variables—then a graph plotted in this way will always be a straight line. For this reason a relation of the first degree in the variables is called a *linear*

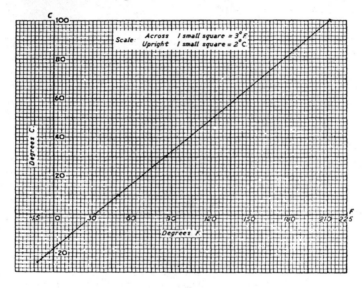

FIG. 120

relation. Thus $2x + 3y = 5$ or $E = \frac{1}{12}L + 4$ will give straight-line graphs; but $y = x^2$ or $pv = 16$ will not.

We have thus seen that a relation between two quantities may be expressed in three distinct ways:

    (i) as a formula connecting the quantities, the quantities being replaced by letters;

    (ii) numerically in tabular form; and

    (iii) pictorially as a graph.

Each has its own special uses and advantages.

THE GENERAL EQUATION OF THE FIRST DEGREE

(This section may be omitted by readers not especially interested in theory.)

Consider the equation $y = mx + c$.
If m and c are regarded as constants this equation is of the first degree in x and y, and so represents a straight line.

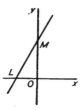

FIG. 121

In order to draw a straight line it is sufficient to know two points on it—in terms of graphs, to know the co-ordinates of two points on it. If these co-ordinates are substituted in the equation to the straight line we shall have two simultaneous equations which on solving will give the values of m and c.

Returning to the straight line $y = mx + c$, we can find where it meets the y-axis by putting $x = 0$ in the equation. This gives $y = c$; that is, in Fig. 121, $OM = c$.

We can find where it meets the x-axis by putting $y = 0$ in the equation. This give $mx + c = 0$, or $mx = -c$; that is, $x = -\dfrac{c}{m}$. So, remembering that OL is measured in the negative direction, we have:

$$LO = \frac{c}{m}.$$

$$\text{Now } \tan M\hat{L}O = \frac{OM}{LO} \text{ (p. 193)}$$
$$= c \div \frac{c}{m} = m.$$

So, in the equation $y = mx + c$, m is the tan of the angle which the straight line makes with the x-axis, and c is the length (measured from the origin) cut off from the y-axis by the straight line. We can think of m as

247

being a measure of its slope or rotation from the x-axis position, c that of its 'shift' in the direction of the y-axis away from the origin. Hence, by just looking at the equation to a straight line, a trained mathematician can tell its position in the x, y plane.

A GRAPH OF AN EQUATION OF THE SECOND DEGREE

We now give an example of the graph arising from an equation of the second degree—that is, an equation which includes no power of a variable higher than the second and no product of the variables other than that of first powers.

The height (h) in feet to which a shell fired by an anti-aircraft gun for a certain vertical velocity of discharge is related to the time (t) in seconds from the moment of firing in accordance with the formula:

$$h = 1000t - 16t^2.$$

Draw the graph of the relation between h and t for intervals of 10 sec. up to $t = 70$ sec.

The calculation of the values of h corresponding to given values of t is best performed, and the results tabulated, as follows:

t	0	10	20	30	40	50	60	70
$1000t$	0	10,000	20,000	30,000	40,000	50,000	60,000	70,000
$16t^2$	0	1,600	6,400	14,400	25,600	40,000	57,600	78,400
h	0	8,400	13,600	15,600	14,400	10,000	2,400	−8,400

The h-row is here found by subtracting the terms in the third row from the corresponding terms in the second row.

We are not concerned with negative values of t, so in drawing the axes we allow for positive values only of t, and for values of h from, say, $-10,000$ to about 16,000 ft. The points are joined by a smooth curve; for this, the curve must rise slightly higher between $t = 30$ and $t = 40$ than the values calculated for $t = 30$, $t = 40$.

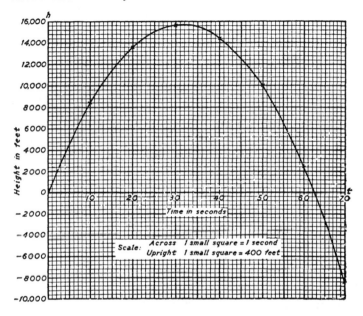

FIG. 122

Deductions from the graph. 1. The highest point reached by the shell—that is, the maximum value of h—is about 15,625 ft. after a time of $31\frac{1}{4}$ seconds.

2. If the shell does not explode it will return to the level through the gun-site ($h = 0$) after $62\frac{1}{2}$ seconds (which is twice the time to the highest point).

3. If it is possible for the shell (supposed unexploded) to fall lower than the level from which it is fired (as from a mountainside or the top of a cliff) the negative values of h give its depth below the level of firing for values of t greater than $62\frac{1}{2}$ seconds.

249

4. If it is set to explode after, say, 25 seconds it will then be 15,000 ft. above gun-site level.

5. It will reach a height of, say, 10,000 ft. after $12\frac{1}{2}$ seconds.

6. If the shell does not explode, then up to $62\frac{1}{2}$ seconds there will be two values for the time for each given height, except in the case of the maximum height, when there is only one value for the time.

7. The rate of climb of the shell becomes less as the time increases from 0 to $31\frac{1}{4}$ seconds. Of course, it is zero at the maximum height.

This is really a kind of 'travel' graph. In mechanics we should call it a 'distance-time' curve.

GRADIENTS

Suppose A, B, C are the points on the curve corresponding to times 10, 20, 30 seconds. Then the average rate of climb

$$\text{between } O \text{ and } A = \frac{8400}{10} = 840 \text{ ft. per sec.,}$$

$$\text{and between } A \text{ and } B = \frac{13,600 - 8400}{10} = 520 \text{ ft. per sec.,}$$

$$\text{and between } B \text{ and } C = \frac{15,600 - 13,600}{10} = 200 \text{ ft. per sec.}$$

But these rates of climb are also the slopes, or *gradients*, of the straight lines OA, AB, BC with respect to the horizontal. So in this distance-time curve the slope, or gradient, of the line joining two points on it gives the average speed (here, rate of climb) between the two times represented by the points. The smaller the time interval is made, the more nearly does the average speed in that time approximate to the actual speed at any point of time in it. Now, when two points on a curve are brought closer until they eventually coincide with each other then, except in very exceptional cases, the straight line joining them will become a tangent to (that is, will touch) the curve; and it will still have a definite slope or gradient. In the case of a distance-time curve this will represent the actual speed at the point.

For a curve in which speed is plotted against time the gradient of the straight line joining two points on it will give the average rate of change of speed—that is, the average acceleration—and the slope of the tangent at a point will give the actual acceleration at that point.

The gradient principle holds generally. If, for instance, y and x are connected by a relation and the corresponding graph is drawn, then the gradient with respect to the x-axis of the tangent to the curve gives the rate at which y is increasing (positive gradient) or decreasing (negative gradient) with respect to x. If the slope is zero (the tangent will then be parallel to the x-axis) y will have a maximum or a minimum value at that point.

In wireless theory the mutual conductance of a valve is the number of milliamps change in anode current per change of one volt in the grid voltage. If a graph is drawn in which current in milliamps is plotted against grid voltage the mutual conductance at any point will be the gradient of the tangent at that point with respect to the voltage axis.

EXERCISE 34

1. If $v = 10 + 32t$ plot a graph showing how v varies when t takes values from 0 to 5. ($t = 0, 1, 2, 3, 4, 5.$)

2. Do the same for $v = 100 - 16t$ for values of t from 0 to 5.

3. In a machine the load (L lb.) raised by an effort (E lb.) is given by the relation $L = 8E - 20$. Draw a graph showing the loads raised by efforts of 5, 10, 15, 20 lb., and use it to find (i) the load raised by an effort of 12 lb., (ii) the effort necessary for a load of 1 cwt.

4. Two quantities x and y are connected by the relation
$$8x + 3y = 50.$$
Draw a graph showing how y varies for values of x from 0 to 7.
Hint: first write the relation as, $y = \frac{1}{3}$ (. . .).)

5. Do the same for $3x + 2y = 4$ for values of x from -2 to 4.

6. Plot the following corresponding values of L and E in a graph:

L	0	10	20	30	40
E	1·5	3·9	4·5	5·7	6·7

There is reason to believe that one pair of the given values is badly wrong. Find which it probably is.

7. In a certain machine it is found that efforts (E) of 20 and 24 lb. weight are needed to move loads (L) of 56 and 112 lb. weight respectively. Assuming that the relation between L and E may be represented by a straight-line graph, draw the graph, and find from it (i) the effort necessary to move 224 lb. weight, (ii) the load that an effort of 30 lb. weight will move, (iii) how much of the effort is wasted on the parts of the machine itself.

8. Draw the graph of $y = x^2$ for values of x from 0 to 6. Use the graph to find (i) $4\cdot5^2$, (ii) $\sqrt{30}$ (1 dec. place).

9. Draw the graph of $y = x^3$ for values of x from 0 to 5. Use the graph to find (i) the cube of $3\cdot6$, (ii) the cube root of 100.

10. If a ball is thrown vertically in the air with a certain velocity it rises s ft. in t sec. where $s = 80t - 16t^2$. Draw a graph showing how s varies when t takes values from 0 to 5, and (i) use your graph to find the greatest height reached by the ball. By finding the gradient at those points calculate its velocity (ii) after 2 sec., (iii) when it is 64 ft. up and descending.

11. A car is bought for £360, and a third of its value at the beginning of each year is set off as depreciation at the end of that year. Draw a graph to show the value of the car at the end of each year up to five years, and from it find (a) the approximate value at the end of $2\frac{1}{2}$ years, (b) when its value will be approximately £100.

12. The following values for the lift of a certain type of aeroplane wing were found by experiment with a model at various air speeds.

Lift in lb. weight	0	5,000	12,800	23,500	28,800	45,000
Velocity in m.p.h.	0	50	80	100	120	150

One of these values for the lift is incorrect. By drawing a graph find which it is.

WHAT IS A FUNCTION?

We have now for some time been dealing with the idea of dependence—F on C, h on t, and so forth. This idea runs like a thread through the whole of the higher mathematics, so before leaving the subject of dependence we shall put the matter a little more precisely.

When two variables are so related that the value of one of them depends upon the value of the other, then the first variable is said to be a function of the second variable.

We can now say that the temperature in degrees Fahrenheit (F) is a function of the same temperature expressed in degrees Centigrade (C), and that the height (h) to which a shell rises is a function of the time (t). In the same way the area of a square (say, A) is a function of the length of the side (say, l), the functional relation being $A = l^2$. In all these cases cited the letter whose value is first changed (as t in the height-time function) is called an *independent variable*; the variable which assumes different values consequent upon changes in the independent variable is called the *dependent variable*.

Of course, one variable may be a function of several other variables—as, for instance, when we say that the volume (v) of a cuboid is given by the formula $V = abc$ where a, b, c are the lengths of its edges. Here a, b, c

are independent variables; V is the dependent variable. What happens if we write this in the form $c = \dfrac{V}{ab}$?

To leave the purely mathematical aspect of the notion for a moment: The speed of a railway train is a function of . . . what? A little consideration will show that it is a function of the steam pressure utilized, the gradient of the track, the state of the rails, the diameter of the piston, the weight of the train, and the number of revolutions made by the driving wheel per minute—to mention some possibilities. It is *not* a function of the scenery or any passenger's anxiety to get on (not in England, at any rate). This seems so obvious and silly that you may wonder why it is stated at all; yet many blunders made by ordinary people without any understanding of mathematical methods are caused by their inability to see what depends on which. The lighting of a fire indoors, for example, is *not* a function of the sunshine outside; yet there is a common, though declining, belief among uneducated people that the sun puts the fire out.

Of course, when the number and nature of the variables in a function have been thought out, only the first stage of mathematical reasoning has been passed. The next step is to decide in what proportion each variable occurs (and whether the proportion is direct or inverse). This can be done by experiment, but we must arrange this so that only one independent variable is allowed to change. Then, by observation and graphical methods, we can see whether the dependent variable varies directly or inversely with the independent variable squared, or cubed, or under a square root, or subjected to any other mathematical operation. After this we can arrange things for the independent variable already examined to be kept constant and the experiment repeated with one of the other variables changing. We go on doing such experiments until we finally arrive at the complete formula involving all the variables. This is the fundamental way of tackling all *new* problems when we have no clue to start with.

You may think that this sounds too simple and that trained mathematicians would not rely on such methods. To some extent this is so, for there are many mathematical techniques; but the simple method above has been used often by eminent scientists. Georg Ohm, for example, spent some years in careful experiment on electrical conduction, keeping everything constant except the voltage supplied and measuring the current, then keeping everything constant except the length of the conductor, and so on. At length he was able to formulate his Law of Conduction—namely, that the electric current depends directly on the supplied voltage, directly on the area of the conductor, inversely on the length of the conductor, and inversely on a special property called the resistivity. Ohm's Law has now been generally accepted and used for more than a hundred years, and for direct-current circuits it has never been proved wrong.

It will be seen that the essence of the method is clear thinking and careful, continued experiment. For instance, what causes a common cold? Each reader will have his answer pat—draughts, wet feet, lack of fresh air, too much fresh air, change of temperature, and so on. Yet are you sure? Haven't you generalized rather quickly and easily? Were you always in the same clothes when you were in the draught that you say caused your cold? Was it always the same time of the year? Were you always alone? If not, were there many people near you? Did any of these have colds? Was it raining? Was it hot inside and cold outside? Had you been travelling just previously in a crowded vehicle, or sitting in a crowded hall? All these queries show that there are very many variables, and that the common cold may be a function of all of them or only some of them.

In this chapter we have seen that if there is a functional relationship of the first or second degrees between two variables that relationship may be depicted graphically by a straight or curved line. For relationships of higher degree it may be possible to draw a curve; in fact the drawing of such a curve, to a mathematician, is often a source of sheer delight.

To sum up, in the applications of mathematics data may be presented in three different ways:

(a) statistical data may be collected from practical experiments by, for example, the engineer or the scientist, and may be exhibited in tabular form; or

(b) they may be illustrated conveniently by means of a curve (and this includes a straight line); and

(c) theoretical considerations may lead to the formulating of a functional relationship.

The reader will readily appreciate that if (c) is given it is possible to deduce (a) or (b). Similarly, given (a), it is possible to obtain (b). The reverse processes—that is to say, given (a) or (b)—to infer a functional relationship may often be a matter of very considerable difficulty. It has often happened that advances in theoretical mathematics have been made by this very means. In such ways general laws, in science perhaps more than in mathematics, have been formulated. We shall refer later to this topic.

Functions may be *single-valued*, as in the case of $y = x^2$, or *many-valued*, as in the case of $x = \pm\sqrt{y}$ (two values here of x for each value of y). They may be *explicit*, as in the case again of $y = x^2$, where a value given to x immediately gives a corresponding value for y; or they may be *implicit*, as in the case of $x^3 + y^3 + 2xy = 1$, where it is not possible to calculate y directly in terms of x.

But these are matters culled from the higher mathematics, and are only given here to indicate to the reader the opening of one of the most modern branches of higher mathematics.

ADDITIONAL TRIGONOMETRICAL NOTE

In Chapter XIV the chief trigonometrical ratios were defined for acute angles only. We are now in a position to define those ratios for angles of any magnitude.

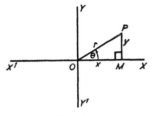

FIG. 123

Let $X'OX$, $Y'OY$ (Fig. 123) be two perpendicular axes. Then if the co-ordinates of P with respect to these axes are (x, y), and $OP = r$, in the case of an acute angle θ, we may rewrite our definitions as:

$$\sin \theta = \frac{y}{r}, \cos \theta = \frac{x}{r}, \tan \theta = \frac{y}{x}.$$

To extend these definitions to an angle of any magnitude all that we have to do is to say that these definitions hold for all positions of P in the four quadrants into which the axes divide the plane, x and y having the signs appropriate to each quadrant and r being regarded as 'signless.'

Numbering the quadrants in the anti-clockwise order, we have, since x is positive in the first and fourth quadrants and negative in the second and third quadrants:

cos θ is positive for an angle between 0° and 90° and for an angle between 270° and 360°; and

cos θ is negative for an angle between 90° and 270°.

R   257

And, since y is positive in the first two quadrants and negative in the third and fourth quadrants:

sin θ is positive for an angle between 0° and 180°; and

sin θ is negative for an angle between 180° and 360°;

tan θ is positive for an angle between 0° and 90° and for an angle between 180° and 270°; and

tan θ is negative for an angle between 90° and 180° and for an angle between 270° and 360°.

It has already been pointed out (see p. 203) that the trigonometrical ratios for angles greater than 360° are repetitions of those for angles between 0° and 360°.

Produce PO to P_2 so that $OP_2 = OP = r$.

Draw P_1OP_3 so that $X'\hat{O}P_1 = \theta$, $OP_1 = OP_3 = r$. Join PP_3, P_1P_2.

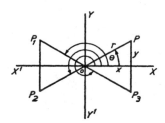

FIG 24

It is easy to prove, by congruent triangles, that PP_3, P_1P_2 are each perpendicular to $X'OX$, and it is clear that $X'\hat{O}P_2$, $X\hat{O}P_3$ are each equal to θ.

So $X\hat{O}P_1 = 180° - \theta$, $X\hat{O}P_2 = 180° + \theta$, $X\hat{O}P_3 = 360° - \theta$. (The angles are measured in the direction of the 'arrowed' arcs of Fig. 124.) It follows that:

$$\sin (180° - \theta) = \sin X\hat{O}P_1 = \frac{+y}{r} = \sin \theta;$$

258

$$\sin(180° + \theta) = \sin X\hat{O}P_2 = \frac{-y}{r} = -\sin\theta; \text{ and}$$

$$\sin(360° - \theta) = \sin X\hat{O}P_3 = \frac{-y}{r} = -\sin\theta.$$

The reader should check for himself that:

cos $(180° - \theta) = -\cos\theta$, cos $(180° + \theta) = -\cos\theta$,
cos $(360° - \theta) = \cos\theta$;

and that:

tan $(180° - \theta) = -\tan\theta$, tan $(180° + \theta) = \tan\theta$,
tan $(360° - \theta) = -\tan\theta$.

Thus sin $140° = \sin(180° - 40°) = \sin 40°$;
cos $200° = \cos(180° + 20°) = -\cos 20°$; and
tan $300° = \tan(360° - 60°) = -\tan 60°$.

PART IV

ARGUMENTS

CHAPTER XIX

DEMONSTRATIVE GEOMETRY

WHO WAS EUCLID?

(Before starting this section, the reader is advised to go over Chapter IX again, particularly pp. 132 to 134.)

WE have already noted that geometry started by being practical and that the Greeks intellectualized it. Some-

EUCLID

where around the year 300 B.C. a man was born who, if we are to judge by the effects of his life on mankind, stands among the great figures of all time. This was a Greek named Eucleides (or, by us, Euclid), who went to Alexandria and became a teacher of geometry. He gathered together what was then known about geometry, and taught his pupils how to demonstrate—that is, show by reasoning—the truth of certain conclusions concerning lines, angles, triangles, parallelograms, circles, and the like.

Euclid arranged his materials in a treatise on demonstrative, or theoretical, geometry which he called the *Elements*. This has passed through more than 2000 editions, and has been for more than twenty-two centuries the encouragement and guide of scientific thought. Most people above the age of fifty who learnt theoretical geometry at school learnt it from books that had changed little from Euclid's own collection. To possess such vitality the *Elements* must have been a very great book on a very great theme.

Some fifty years ago geometry books began to be produced which took a great deal of the inflexibility out of Euclid's system. Nowadays the presentation is made even more attractive; yet the logical lucidity of Euclid's work still shines out from every page.

THE STARTING-POINT

Whether one is dealing with mathematics or with other things, we have to be sure that the reasons given to justify statements are acceptable ones. In the Law Courts, for instance, there are rules for giving evidence, and every person acting in a judicial capacity is guided, and to a certain extent bound, by judgments given in earlier cases of a like nature. It is not many years since the evidence afforded by finger-prints was first admitted; it may be that some day the evidence of a 'lie-detector' will be allowed. But our point here is that evidence not approved is ruled out.

In mathematics, too, reasons given for statements must be acceptable ones. Reasons which seem perfectly good to individuals—such as saying that straight lines are parallel because they *look* parallel (are *AB* and *CD*, in Fig. 125, parallel?) or that angles are equal because *on measurement* they appear to be so—are not necessarily good

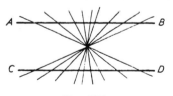

Fig. 125

reasons. There are certain rules of correct thinking which must be followed.

Now there are some things which common sense tells everybody are true. No one doubts that two apples added to three apples make five apples; nor would anyone seriously question that things that are equal to the same thing are equal to one another.

We have to make this common-sense approach to geometry at the beginning. Euclid did it by assuming the existence of certain self-evident truths. They were called *axioms* and *postulates*; axioms had to do with matters theoretical, postulates with constructions.

AXIOMS AND POSTULATES

Euclid's axioms—he called them 'common notions'—were such statements as: "Things that are equal to the same thing are equal to one another"; "If equals be added to equals the results are equal"; "Things that are double the same thing or equal things are themselves equal"; and "The whole is greater than its part." In addition, there were the more particularly geometrical axioms: "Magnitudes that can be made to coincide with each other are equal"; "Two straight lines cannot enclose a space"; and "All right angles are equal." There was also an axiom on which his theory of parallels was based. This proved so unsatisfactory in the course of time that various substitutes were proposed for it; that in general use now was suggested in 1818 by John Playfair, Professor of Mathematics at Edinburgh University: "Two intersecting straight lines cannot both be parallel to a third straight line."

Euclid's *postulates* were: "Let it be granted that a straight line may be drawn from any point to any other point; that a finite—that is to say, terminated—straight line may be produced to any length in that straight line;

that a circle may be described from any centre, at any distance from that centre—that is, with a radius equal to any finite straight line drawn from the centre."

The reader will surely say, "But all this is so very obvious." This may be true enough; but for any project a foundation has to be laid, and the axioms and postulates are the very foundation stones on which the mighty edifice of logical geometry has been built up. For, however we may once have disliked learning geometrical theorems, it cannot be gainsaid that as a training in ordered thinking and reasoning there is nothing to beat theoretical geometry.

Starting with the axioms and postulates, and proceeding via the *definitions*—many of which have already appeared in the earlier pages of this book—Euclid built up, step by step, his *Elements*.

GEOMETRY TO DAY

After laying his foundation stones of axioms and postulates and stating his definitions, Euclid made use of two main ideas in building his superstructure: parallels and congruence.

We have already mentioned Playfair's axiom on parallels on which the parallel theory of fifty years ago rested. Nowadays what Euclid *proved* about parallels is generally *assumed* —namely (for convenience of description we shall use the letters of Fig. 126), if *AB* and *CD* are parallel and a straight line crosses them, then:

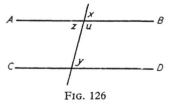

FIG. 126

$x = y$ (they are called *corresponding angles*); from which it follows since $x = z$ that

263

$y = z$ (they are called *alternate angles*),
and $u + y =$ two right angles (they are often called
allied angles), since $u + z =$ two right angles.

Conversely—that is to say, oppositely—if any of these
three statements is true AB and CD are parallel.

With regard to congruent triangles Euclid purported
to *prove* congruence under the conditions mentioned at
pp. 174, 175. Nowadays his proofs are considered unsatisfactory—particularly by the philosophers—and so congruence is assumed for the conditions given at pp. 174, 175.

At the present time matters seem to be advancing still
farther. There are many who hold that the idea of
similarity (pp. 171–173) is fundamental, and that congruence should be developed as a particular case of
similarity. Similarity can also be made a basis from
which parallel theory may be deduced. Some day we
may see an elementary course of geometry built up from
this concept; but that day is not yet.

WHAT IS A PROPOSITION?

Euclid called his geometrical arguments *propositions*.
There are two classes of propositions: theorems and
problems (or constructions).

A *theorem* is a proposition in which some geometrical
statement has to be proved.

A *problem* (or construction) is a proposition in which
some particular figure or line(s) has to be drawn.

The following are four examples of propositions:
The opposite sides of a parallelogram are equal.
The angle in a semicircle is a right angle.
To divide an angle into two equal parts.
To construct the tangents to a circle from an external
    point.

The first two of these are theorems, and the other two

are problems. Notice that a problem here does not mean quite the same thing as what is popularly known as a 'teaser.'

A proposition is divided into four parts:

(i) The *general enunciation*, in which a general statement is made of a theorem or problem.

(ii) The *particular enunciation*, which is the special application of (i) to the figure which has been drawn and lettered.

(iii) The *construction*, which is the drawing of any additional lines or circles that may be necessary.

(iv) The actual *proof*, of the statement in the case of a theorem, of the validity of the construction in the case of a problem.

In (i) and (ii) we have another instance of the principle of abstraction, to which we have already referred. A carpenter might ask us to *prove* that the distances between opposite corners of a rectangular framework are equal (see p. 187)—that is (i); for our proof we should have to make a rough sketch of a rectangle, letter its corners, and name the diagonals to be proved equal—that is (ii).

The enunciation, general or particular, of a proposition is itself divided into two parts: what is actually given (also called the *hypothesis*), and what is to be proved or constructed. And so, in writing out a proposition of which the general enunciation is given, it is usual to have these headings:

(i) Given; (ii) To Prove; (iii) Construction; (iv) Proof. The words used may vary slightly with different writers.

WHAT MAKES A VALID PROOF?

We have already partly answered this question in the last few pages, but before we proceed to give some actual examples of proofs the 'rules' of valid proof are here gathered together. It must be remembered that a proof is an argument, each step of which must be supported by a valid reason; where a reason cannot be found think twice before making a statement in geometry!

These reasons may be either:

(i) Euclid's axioms, and the other 'agreed' axioms, such as those relating to the conditions of congruence of triangles and the angles made by parallels with a crossing line;

(ii) the facts that we are given in the enunciation of the proposition;

(iii) any proposition which has already been proved to be true; and

(iv) any property of the figure under consideration which is a matter of definition—*e.g.*, a parallelogram is a quadrilateral whose opposite sides are parallel.

It is usual to place the letters 'Q.E.D.' at the end of a proof. They are the initial letters of the Latin words *Quod erat demonstrandum*, meaning 'which was to be proved.'

REAL AND APPARENT PROOF

Most people are familiar with these lines from Shakespeare:

> To thine own self be true
> And it must follow, as the night the day,
> Thou canst not then be false to any man.

Let us examine this excerpt from the point of view of rigid proof. It would appear that this is an argument based on the observation that night always follows day.

Night certainly follows day; this is undoubtedly an axiom.

But from this Polonius goes on to say that if you are true to yourself you cannot be false to anyone. Is there an axiom implied here? No. The argument is by analogy. There is no connexion between truth to oneself and falseness to another man which can be based upon the fact that night follows day.

This seeming irrelevance has its counterpart in geometrical proof. Many a student has written in substance, "Since $x = y$, therefore $p = q$." Often the premiss, $x = y$, is correct; the conclusion, $p = q$, on the other hand, may bear no relation to the premiss.

SPECIMEN PROPOSITIONS

(*a*) The sum of the three angles of a triangle is two right angles (p. 139).

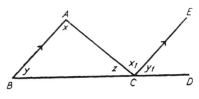

FIG. 127

Given.   A triangle ABC.

To prove. $x + y + z = 2$ right angles.

Construction. Produce BC to any point D.
                     Draw CE parallel to BA.

Proof. $x = x_1$ (alternate angles; CE parallel to BA);
                     $y = y_1$ (corresponding angles; CE parallel to BA);
          and $z = z$.

267

Adding, $x + y + z = x_1 + y_1 + z$
$$= 2 \text{ right angles, since } BCD \text{ is a}$$
straight line. Q.E.D.

Queries.   (i) What axioms are used here?
           (ii) What postulates are used here?
           (iii) What 'agreed axioms' are used here?

Notes.   We have also: $x_1 + y_1 = x + y$,

that is, $A\hat{C}D = \hat{A} + \hat{B}$.

This is called a *corollary* to the original proposition. A corollary to a proposition is an additional fact which may be deduced from the demonstration of the theorem.

This proof should be compared with the checking of the same property at p. 139.

(*b*) If two sides of a triangle are equal then the angles opposite them are equal.

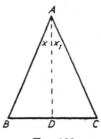

Fig. 128

Given.   A triangle ABC with $AB = AC$.

To prove. $\hat{B} = \hat{C}$.

Construction. Draw AD to bisect $B\hat{A}C$. Let it meet BC
              at D.

Proof.   In the triangles BAD, CAD
               $AB = AC$ (given);
                 $x = x_1$ (construction);
               $AD = AD$.

Therefore the triangles are congruent (two sides, included angle); and so, in particular,

$$\hat{B} = \hat{C}.$$ Q.E.D.

Queries. (i) What 'agreed axiom' is used here?

(ii) Examine Euclid's postulates (given at p. 262). Do you think that you are assuming the existence of another postulate here? What is it?

Notes. Since we have proved that the triangles *BAD*, *CAD* are congruent—that is, equal in all respects—it also follows that:

$$BD = DC; \text{ and that}$$
$$A\hat{D}B = A\hat{D}C.$$

So, by the definition of a right angle (p. 136), *AD* is perpendicular to *BC*.

We might thus have extended the general enunciation of the theorem by saying: if two sides of a triangle are equal then the bisector of the angle between them bisects the third side (or base) of the triangle and is perpendicular to the base.

(*c*) If two points on the circumference of a circle are joined to the centre of the circle, and also to any point on the circumference, then the angle made at the centre is double that made at the circumference.

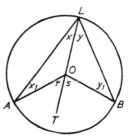

FIG. 129

Given. Two points *A* and *B* on the circumference of a circle of centre *O*; *L* is any other point on the circumference. *AO*, *OB*, *AL*, *LB* are joined.

To prove. $A\hat{O}B = 2 A\hat{L}B.$

269

Construction. Join LO, and produce it to T.

Proof. $\qquad x = x_1$ (proved in proposition b).

$\qquad$ But $\quad r = x + x_1$ (corollary to proposition a).

$\qquad$ So $\quad r = 2x$.

$\qquad$ Similarly, $s = 2y$.

$\qquad$ Adding, $r + s = 2x + 2y = 2(x + y)$.

That is, $A\hat{O}B = 2\ A\hat{L}B$. $\qquad\qquad$ Q.E.D.

Queries. As for proposition (a).

Also: do you see how, in proving this proposition, use has been made of propositions (a) and (b)?

Notes. In certain positions of L, where AL or BL cuts OB or OA respectively, this proof does not hold exactly as it stands; it is then necessary to prove that

$$r - s = 2(x - y),$$

and the same result is obtained. Try drawing the figure yourself.

A number of important propositions follow from this proposition. Perhaps the reader will be able to establish them for himself from the hints given below.

(i) The angle in a semicircle is a right angle:

$$x = 2x_1 \text{ (Fig. 130)}.$$

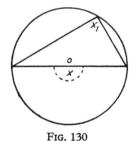

FIG. 130

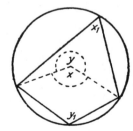

FIG. 131

(ii) If a quadrilateral has its corners on the circumference of a circle (that is, is inscribed in a circle) then each pair of opposite angles adds up to two right angles:

$$x = 2x_1, \quad y = 2y_1 \text{ (Fig. 131)}.$$

(iii) $A\hat{L}B = A\hat{M}B$ (Fig. 132).

270

(*d*) To bisect a given angle—that is, to divide it into two equal parts.

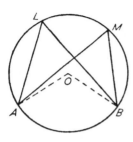

FIG. 132

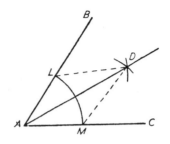

FIG. 133

Given. An angle *BAC*.

To construct. A straight line bisecting $B\hat{A}C$.

Construction. With centre *A* and any convenient radius draw an arc of a circle to cut *AB* and *AC* at *L* and *M*.

With centres *L* and *M* and a radius greater than $\frac{1}{2}LM$ draw two (equal) circles to cut at *D*.

Join *AD*.

Then *AD* bisects $B\hat{A}C$.

Proof. Join *LD, MD*.

In the triangles *ALD, AMD*

$AL = AM$ (equal radii);

$LD = MD$ (equal radii); and

$AD = AD$.

Therefore the triangles are congruent (three sides); and so, in particular:

$$L\hat{A}D = M\hat{A}D;$$

that is, $B\hat{A}D = C\hat{A}D$.

So *AD* bisects $B\hat{A}C$.

271

Notes. (i) When we draw two circles to cut at D it is only necessary to draw the portions of the two circles in the neighbourhood of the place where they may be expected to meet.

(ii) Another method, not dependent upon the use of compasses, is given at p. 189.

THE CONVERSE OF A THEOREM

In a theorem if we are *given* certain facts it may be possible to *prove* certain conclusions.

The converse of a theorem is obtained by interchanging in its statement the given facts and the conclusions.

The converses of *some* theorems are true. For example, we know that if two sides of a triangle are equal (proved at p. 268) the angles opposite them are equal. The converse of this can also be proved to be true— namely, that if two angles of a triangle are equal then the sides opposite them are equal.

The converses of some theorems are definitely not true. For example, it can be proved that the diagonals of a rectangle are equal; but the converse is not necessarily true—namely, that if the diagonals of a quadrilateral are equal the quadrilateral is a rectangle. For we have only to draw two intersecting lines of equal length and join their ends; clearly their joins do not form a rectangle in every possible position of the two equal lines.

Thus we have to prove the converse of a theorem before we can make use of it in a proof.

PART V

MAKING THE WORK EASIER

CHAPTER XX

CALCULATING DEVICES

GRAPHS

WE shall here briefly remind the reader of some of the graphs dealt with in Chapter VIII. It was found there that squared paper could be used to advantage in shortening calculations. Instances cited were the reckoning of selling prices from cost prices, and *vice versa*, for particular rates of profit; graphs for calculating discount; 'conversion' graphs, for changing from one system of units to another, such as miles to kilometres and *vice versa*; graphs which enable products and quotients for a particular multiplier or divisor to be read off immediately; and—a useful extension of this latter kind—graphs for estimating speeds or distances for particular rates of travel.

The reader will realize that the possibilities of this type of graph are endless. The accuracy of any calculation done in this way will depend upon the scale and size of the graph; over a limited range it may well be sufficient for all practical purposes. A great disadvantage, however, is that for each operation a separate graph is necessary. Yet surprisingly great use is made of this kind of graph in all walks of life.

READY RECKONERS

There was a time when the use of ready reckoners was rather despised. The professional mathematician rather looked down his nose at them (though he readily made use of other kinds of tables for the simplification of calculations), and encouraged the same attitude in his pupils. Nowadays ideas are changing; a great amount of purely mechanical manipulation is, of course, still absolutely necessary in order to impress and underline principles. But any aid to more speedy (and possibly more accurate!) computation is not frowned upon as formerly, and the tendency now is to teach the intelligent use of ready reckoners and other calculation devices.

After all, there is no particular merit in spending a long time on a numerical operation when the answer may be read at once from a table. And there is really no reason why a ready reckoner should be kept, almost shamefacedly, 'under the counter.'

It is not proposed to consider the various kinds of ready reckoners in this book; anyone familiar with the use of any kind of table will readily adapt himself to the manipulation of a ready reckoner.

MORE ABOUT EXPONENTS

We saw at p. 31 that:

(i) $a^m \times a^n = a^{m+n}$, and
(ii) $a^m \div a^n = a^{m-n}$,

where m and n are whole numbers and m is greater than n; and also that:

$$a^0 = 1.$$

In the same way:

$$(a^2)^3 = (a \times a) \times (a \times a) \times (a \times a) = a^6.$$

This is a particular case of another general rule:

$$(a^m)^n = a^{mn}.$$

We have checked the truth of these rules for positive whole-number values of m and n (and (ii) above only for values of m greater than n). In what follows we shall assume that they hold for all values of m and n, positive and negative, whole numbers and fractions. This can be proved, but the proof is not really necessary for our purpose.

Next we must find meanings for negative and fractional powers—that is, for expressions like a^{-n} and $a^{\frac{1}{2}}$.

If we put $m = 0$ in (ii) we have:

$$a^0 \div a^n = a^{0-n};$$

and since $a^0 = 1$, and $0 - n = -n$, this gives:

$$1 \div a^n = a^{-n};$$

that is,

$$a^{-n} = 1 \div a^n = \frac{1}{a^n}.$$

Finally, if we put $m = n = \frac{1}{2}$ in (i) we have:

$$a^{\frac{1}{2}} \times a^{\frac{1}{2}} = a^{\frac{1}{2}+\frac{1}{2}} = a^1 = a,$$

and so

$$a^{\frac{1}{2}} = \sqrt{a}.$$

Similarly

$$a^{\frac{1}{3}} = a \text{ cube root of } a; \text{ and so on.}$$

LOGARITHMS

Suppose $a = 10$; then

$$10^m \times 10^n = 10^{m+n};$$
$$10^m \div 10^n = 10^{m-n};$$
$$(10^m)^n = 10^{mn};$$
$$10^{-n} = \frac{1}{10^n};$$
$$10^{\frac{1}{2}} = \sqrt{10};$$

and so on.

If we can express numbers as powers of 10 (or, indeed, of any other base) we can replace multiplication by addition of indices, division by subtraction of indices, and obtain powers and roots by multiplication and division of indices.

These indices which express numbers as powers of 10 are called the common logarithms, or just the *logarithms*, of the numbers to *base* 10. The logarithm of x is written, for short, log x.

THE LOGARITHMIC FORM OF THE RULES

Suppose $10^m = x$, $10^n = y$, so $xy = 10^m \times 10^n = 10^{m+n}$. Then, by our definition of a logarithm:

$$m = \log x, \quad n = \log y, \quad m + n = \log xy.$$

But $m + n = \log x + \log y$, also.

$$\text{So } \log xy = \log x + \log y \qquad . \qquad . \qquad \text{(I)}$$

$$\text{Similarly} \qquad \log \frac{x}{y} = \log x - \log y \qquad . \qquad . \qquad \text{(II)}$$

In the same way we may show that:

$$\log x^5 = 5 \log x; \quad \log \sqrt{x} = \log x^{\frac{1}{2}} = \tfrac{1}{2} \log x;$$

and so on.

LOGARITHM TABLES

It is possible to *calculate* or to find from a graph the power of 10 which is equal to a given number. This is a tedious business, and to save time tables have been prepared from which the log of a number to base 10 may be read. It would be impossible to construct tables giving the logs of all numbers great and small, so they are generally constructed for numbers of four, five, or seven figures. Those mostly used are for numbers of four figures, the figures being counted from the first figure on the left of the number which is not a zero. (This covers decimal fractions.) So in using four-figure tables a number whose logarithm is desired must, if necessary, first be corrected, or rounded off, to four figures.

The invention of logarithms is attributed to a Scotsman, John Napier (1550–1617), Laird of Merchiston, one of the landed aristocracy of Scotland. The first base which he used was not the 10 of common logarithms, but a quantity known as *e* in the higher mathematics. In 1624 Henry Briggs, a professor of geometry in London, completed Napier's task by publishing a table of logs of numbers with the base 10. Recent work by two American archæologists has shown that an Old Babylonian tablet answers the question to what power must a certain number *a* be raised to yield a given number. This

JOHN NAPIER

problem is identical with finding the logarithm to the base *a* of a given number. So in the light of further research it may appear that Napier really rediscovered logarithms.

CHARACTERISTIC AND MANTISSA

The next question is, "How can we use log tables for any number of four figures when regard must be paid to the position of the decimal point?"

Consider the following table; it is built up in a way similar to that given at p. 31 for powers of 2:

Number	...	1000	100	10	1	$\frac{1}{10}$	$\frac{1}{100}$	$\frac{1}{1000}$	....
Number as a power of 10		10^3	10^2	10^1	10^0	10^{-1}	10^{-2}	10^{-3}	
Log of number to base 10		3	2	1	0	−1	−2	−3	

So the log of a number between

100	and 1000	will be	2 +	a decimal fraction	
10	„ 100	„	1 +	„	„
1	„ 10	„	0 +	„	„
·1	„ 1	„	− 1 +	„	„
·01	„ ·1	„	− 2 +	„	„
·001	„ ·01	„	− 3 +	„	„

and so on.

The integral, or whole-number, part of the log is called the *characteristic*; the decimal part is called the *mantissa* (Latin for 'worthless addition' or 'make-weight'!)

From the above we have the following rules for the determination of characteristics:

(i) *If the number is greater than one the characteristic of its log is one less than the number of digits before the decimal point.*

(ii) *If the number is less than one the characteristic of its log is negative, and it is numerically one more than the number of 0's immediately following the decimal point.*

So   log 6·853   = 0 + a decimal fraction
                   = 0·8359 (as we shall see later).

Now log 68·53   = log (10 × 6·853)
                   = log 10 + log 6·853 (p. 276)
                   = 1 + 0·8359.

So   log 68·53   = 1·8359.

Similarly,
       log 685·3   = 2·8359;
       log 6853   = 3·8359;
       log 68530   = 4·8359;

and so on.

Also, $\log 0\cdot6853 = \log(6\cdot853 \div 10)$
$= \log 6\cdot853 - \log 10$ (p. 276)
$= 0\cdot8359 - 1.$

This is usually written:
$\log 0\cdot6853 = \bar{1}\cdot8359,$
and read 'bar one point eight three five nine.'

Similarly,
$\log 0\cdot06853 = \log(6\cdot853 \div 100)$
$= \log 6\cdot853 - \log 100$ (p. 276)
$= 0\cdot8359 - 2 = \bar{2}\cdot8359;$
$\log 0\cdot006853 = \bar{3}\cdot8359;$
and so on.

It thus appears that the mantissæ (plural of mantissa) of all numbers with the same four figures in the same order are the same, the characteristic being altered according to the position of the decimal point. The characteristic is thus a sort of 'decimal-point regulator.'

Note that the mantissa of a log is always positive, but the characteristic may be positive, zero, or negative. Thus $\bar{3}\cdot4526$ means $-3 + \cdot4526$, and not $-3\cdot4526$.

TO FIND THE LOG OF A NUMBER FROM THE TABLES

First 'correct' the number to four figures, introducing 'place-holding' zeros if necessary. Log tables only give the mantissæ; the characteristics are written down according to the rules given above.

Suppose the mantissa is required for the figure 6853. Turn to the log tables at the end of the book. The first two figures (68) are looked for in the extreme left-hand column of each page; the third figure (5) is found by looking along the top row of the page. Place a ruler or a straight edge of paper across and under the 68 row, and follow with the eye down the column headed 5 until the ruler or paper is reached; this gives 8357. Now look up the fourth figure (3) in the row above the columns on the extreme right-hand side of the page (the so-called 'difference' columns); under the 3 is found, in the same row as 68, the figure 2. This 2 is added to 8357, giving 8359.

So   log 6·853   =   ·8359 (we now insert the deci-
                            mal point);

and so log 68·53   = 1·8359;

    log 6853   = 3·8359;

    log 68530000   = 7·8359;

    log 0·6853   = $\bar{1}$·8359;

    log 0·06853   = $\bar{2}$·8359;

    log 0·00006853   = $\bar{5}$·8359

and so on.

If the number has only three figures the 'difference' columns are not used; if the number has only two figures those figures are found in the left-hand column of the table, and the mantissa is looked for under the 0 of the top row. To find the mantissa for a number of one figure only—say, 4—proceed as for 40. The mantissa for 40 is 6021;

so log 40 = 1·6021, log 4 = ·6021, log 400 = 2·6021.

EXERCISE 35

Use the tables to write down the logs of:

4·8, 7·3, 2·5, 6·6, 2·34, 5·39, 4·78, 6·395, 5·348, 5, 6, 7, 20, 50, 45·6, 83·2, 36·5, 345·6, 874·9, 235·3, 46·57, 92·36, 3586, 2983, 123456, 892·35, ·62, ·58, ·562, ·3894, ·9, ·034, ·0683, ·00364, ·0000829, ·0606, ·0015748, ·000000002937.

ANTILOGARITHMS, OR ANTILOGS

Log tables may be used for finding a number whose log is given, by the process of working backward, but usually antilog tables are provided for this.

To find a number the mantissa of whose logarithm is given we follow the process described above for finding the log of a number, but in the antilog tables. The position of the decimal point is indicated by the characteristic. If the characteristic is positive the number of figures before the decimal point will be one more than the numerical value of the characteristic, place-holding zeros being supplied when necessary. If there is no characteristic, or a zero characteristic, the decimal point is placed after the first figure. If the characteristic is negative the number will be less than one, and the number of 0's immediately following the decimal point will be one less than the numerical value of the characteristic. Thus:

antilog ·4657 = 2·922; antilog 2·3407 = 219·2;
antilog 4·5983 = 39660; antilog $\bar{1}$·2538 = ·1794;
antilog $\bar{4}$·6738 = ·0004719.

(Note the use of place-holding zeros.)

EXERCISE 36

Use the antilog tables to write down the antilogs of:
·36, ·58, ·567, ·289, ·3547, ·7598, 1·3469, 2·9475, 3·3333, 5·7003, $\bar{1}$·4749, $\bar{2}$·5607, $\bar{2}$·7356, $\bar{3}$·5647, $\bar{4}$·8625.

CALCULATIONS WITH THE AID OF LOGS

The reader will realize that logs and antilogs to four places must necessarily have a 'corrected' figure in the last place. It is therefore usual, in writing down an answer obtained by using logs, to give it 'correct' to three figures.

(*a*) *Positive characteristics only.*

In our calculations the general method is as follows:

(i) *Multiplication.* Find the logs of the numbers to be multiplied together, add the logs, and find the antilog of the result; (log xy = log x + log y).

For example:

$3 \cdot 476 \times 22 \cdot 63 = 78 \cdot 7$ (correct to 3 figures).

No.	Log
3·476	·5411
22·63	1·3547
78·66	1·8958

The working is done in the framework on the right. 1·8958 is obtained by adding the logs of 3·476 and 22·63, and 78·66 is the antilog of 1·8958.

(ii) *Division.* In this case the logs are subtracted;

$$(\log \frac{x}{y} = \log x - \log y).$$

For example:

$234 \cdot 5 \div 97 \cdot 6 = 2 \cdot 40$ (correct to 3 figures).

No.	Log
234·5	2·3701
97·6	1·9894
2·403	·3807

Here ·3807 is obtained by subtracting log 97·6 from log 234·5, and 2·403 is the antilog of ·3807.

(iii) *Powers.* Here the log of the number is multiplied by the figure indicating the power to which the number is to be raised; ($\log x^5 = 5 \log x$).

For example:

$(17 \cdot 29)^5 = 1{,}540{,}000$ (correct to 3 figures).

No.	Log
17·29	1·2377
(17·29)⁵	6·1885
1,544,000	

The log of 17·29 is multiplied by 5 to give the log of (17 29)⁵.

(iv) *Roots.* In this case the log of the number is divided by the figure indicating the root which is required; (the log of the cube root of $x = \frac{1}{3} \log x$).

For example:

The cube root of $134\cdot5 = 5\cdot12$ (correct to 3 figures).

No.	Log
134·5	2·1287
(134·5)$^{\frac{1}{3}}$	·7096
5·124	

2·1287 is divided by 3 to give the log of the cube root of 134·5. When dividing, the last figure is 'corrected' if necessary.

EXERCISE 37

(Give all answers corrected to three figures.)

1. $3\cdot657 \times 7\cdot38$.
2. $3\cdot58 \times 27\cdot6$.
3. $6\cdot54 \times 237$.
4. $9\cdot673 \times 1\cdot3647$.
5. $4\cdot32 \times 5\cdot73 \times 2\cdot89$.
6. $47\cdot56 \times 59 \times 3\cdot14$.
7. $38\cdot7 \div 3\cdot69$.
8. $45\cdot83 \div 2\cdot46$.
9. $3859 \div 57\cdot82$.
10. $40{,}023 \div 637\cdot5$.
11. $897\cdot5 \div 5\cdot328$.
12. $(4\cdot35)^2$.
13. $(45\cdot6)^3$.
14. $(2\cdot984)^6$.
15. $\sqrt{32\cdot3}$.
16. $\sqrt{165\cdot8}$.
17. Find the cube root of 436·8.
18. Find the fourth root of 79·3.

(b) *Negative characteristics.*

(i) *Multiplication.*

$64\cdot0 \times \cdot09356 = 5\cdot99$ (correct to 3 figures).

No.	Log
64·0	1·8062
·09356	$\bar{2}$·9711
5·988	·7773

On adding the mantissæ there is a carrying figure of 1. We then proceed: $1 + \bar{2} = \bar{1}$; $\bar{1} + 1 = 0$. Remember $\bar{2} = -2$.

(ii) *Division.*

27·65 ÷ 0·0987 = 280 (correct to 3 figures).

No.	Log
27·65	1·4417
·0987	$\overline{2}$·9943
280·2	2·4474

On subtracting the mantissæ 1 has to be 'borrowed'; it is paid back as in arithmetic, giving $1 + \overline{2} = \overline{1}$. To subtract characteristics change the sign of the lower one and add; here $1 - \overline{1} = 1 + 1 = 2$.

(iii) *Powers.*

$(·7624)^3 = ·443$ (correct to 3 figures).

No.	Log
·7624	$\overline{1}$·8822
$(·7624)^3$	$\overline{1}$·6466
·4432	

On multiplying the mantissa by 3 there is a carrying figure of 2; multiply the characteristic by 3, $3 \times \overline{1} = \overline{3}$, add 2, giving $\overline{1}$.

(iv) *Roots.*

The fifth root of ·04563 = ·539 (correct to 3 figures).

No.	Log
·04563	$\overline{2}$·6593
$(·04563)^{\frac{1}{5}}$	$\overline{1}$·7319
·5394	

We cannot divide $\overline{2}$ by 5 without having a negative remainder to carry on to the mantissa. But we get out of the difficulty by writing $\overline{2} = \overline{5} + 3$, and then:

$\overline{2}·6593 \div 5 = (\overline{5} + 3·6593) \div 5 = \overline{1} + ·7319 = \overline{1}·7319.$

(Similarly, $\overline{3}·2586 \div 2 = (\overline{4} + 1·2586) \div 2 = \overline{2}·6293.$)

EXERCISE 38

(Give all answers corrected to three figures.)

1. $24 \times \cdot564$. 2. $\cdot4872 \times \cdot7835$.
3. $164 \times \cdot02345$. 4. $\cdot016 \times \cdot028$.
5. $3\cdot2 \times \cdot0045 \times 1947$. 6. $35\cdot4 \div 48\cdot5$.
7. $383 \div 5638$. 8. $\cdot0384 \div 19$.
9. $14\cdot76 \div 836\cdot5$. 10. $(\cdot356)^2$.
11. $(\cdot0287)^2$. 12. $(\cdot3532)^3$.
13. $\sqrt{\cdot846}$. 14. $\sqrt{\cdot1}$.
15. $\sqrt{\cdot06523}$. 16. Find the cube root of $\cdot366$.
17. Find the cube root of $\cdot8$.
18. Find the tenth root of $\cdot9845$.

MUSIC AND MATHEMATICS

When a pianoforte string is struck by a hammer vibrations are set up in the string. The pitch of a note depends upon the frequency of the vibrations. When middle C is struck, the frequency of the vibrations set up is 256 to 265 per second (according to which pitch is accepted—British Standard Concert pitch makes A = 440). The ratio of the frequencies of two notes is called the interval between the notes. One note is an octave above another note when the ratio of their frequencies is 2 : 1.

In an octave there are twelve semitones. To enable a pianist to play in any key the intervals between semitones must be as nearly as possible the same. (Hence the so-called scale of equal temperament.) And so, if the frequency-ratio between two consecutive semitones is x, then, remembering that the ratio for an octave is 2 : 1, we must have:

$$x^{12} = 2.$$

285

We use logs to find the value of x, for no other way is open to us.

$$\log 2 = \cdot 3010.$$
$$\text{So } \log x = \log 2 \div 12 = \cdot 0251,$$
$$\text{giving } x = 1\cdot 059.$$

So, if we take the frequency of C as 256, the frequency of

$$\text{C} \sharp = 256 \times 1\cdot 059$$
$$= 271.$$

No.	Log
256	2·4082
1·059	·0251
271·2	2·4333

and of

$$G = 256 \times (1\cdot 059)^7$$
$$= 384.$$

No.	Log
256	2·4082
$(1\cdot 059)^7$	·1757
383·6	2·5839

If the frequency of C is taken as 261·6:

$$A = 261\cdot 6 \times (1\cdot 059)^9$$
$$= 440.$$

No.	Log
261·6	2·4176
$(1\cdot 059)^9$	·2259
440	2·6435

The determination of frequencies of the other notes in the octave is left as an exercise for the student.

Note. The above calculations are for a piano or any instrument with fixed notes. Stringed instruments can use the correct diatonic scale, in which the frequency-ratios are not quite the same as with the scale of equal temperament.

CHAPTER XXI

MECHANICAL AIDS FOR CALCULATION

THE SLIDE-RULE

IF we slide one ordinary geometrical rule against another until its zero mark lies opposite the 4-in. mark on the other then the 7-in. mark on the first rule will lie opposite the 11-in. mark on the other. By this mechanical operation the arithmetical calculations $4 + 7 = 11$ and $11 - 7 = 4$ are carried out.

Suppose that instead of having two rules marked into divisions of equal length they are divided and marked in such a way that distances from the left-hand end corresponding to the numbers 1, 2, 3, ... are proportional not to the numbers themselves but to the logarithms of the numbers. Then, from

$$\log x + \log y = \log xy$$

and
$$\log x - \log y = \log \frac{x}{y},$$

it is clear that we can use these two rules for multiplication and division just as ordinary geometrical rules may be used for addition and subtraction. This is the principle of the slide-rule.

The modern slide-rule is, of course, a more elaborate affair. Instead of two rules sliding side by side there is a fixed rule containing a rectangular groove in which the other rule, called the slide, moves. The upper parts of the rule and slide contain two (logarithmically divided) scales exactly alike; the lower parts of the rule and slide each contain only one scale, its length being double that of each of the upper ones. There is also a movable frame, or cursor, which slides along the edges of the slide-rule. On the transparent panel of the cursor a fine hair-line is engraved at right angles to the scales. It enables corresponding points on the fixed and sliding scales to be easily read.

SLIDE-RULE

It is not our purpose here to describe any further how a slide-rule is used; this may be found in the appropriate text-books. With practice quite complicated arithmetical operations can be performed rapidly. Of course, the use of the slide-rule is not limited to multiplication and division; squares and square roots can be read off at once, and with a little manipulation cubes and cube roots can be computed. A good slide-rule also fulfils many of the functions of logarithm and trigonometry tables.

There are other kinds of slide-rules, cylindrical and circular, but the ones described above are most generally used. There are slight differences of detail, but not of principle, between the various common makes of slide-rules.

THE SLIDE-RULE IN HISTORY

The history of the development of the slide-rule is interesting. The first men to make use of the principle were two clergymen, Edmund Gunter (1581–1626) and William Oughtred (1575–1660). Both were country parsons content to stay in their parishes studying, writing, and teaching. Sir Christopher Wren was once a pupil of

Oughtred's. Samuel Pepys, the famous diarist, found the slide-rule "very pretty for all questions of arithmetick."

Sir Isaac Newton, most famous of all British mathematicians, suggested an improvement in 1675, but it was not until 1850 that a French army officer, Amédée Mannheim (1831–1906), standardized the modern slide-rule.

Slide-rules, like ready reckoners, were not popular in the nineteenth century and the early part of this century, partly because it was alleged that they were

WILLIAM OUGHTRED

good only for approximate work, partly on the grounds that mechanical instruments have no place in the study and applications of mathematics. These arguments are not now held so seriously.

EARLY FORMS OF CALCULATION

Early in this book we briefly reviewed the beginnings of arithmetic—how man devised symbols to express numbers, and how he painfully learnt the ways of performing calculations. The abacus, still used in Russia, China, and Japan, was a primitive form of calculating machine, and the counter board, a species of abacus, gave its name to many things in our commercial system.

There was also the 'tally-stick.' When anyone owed money he would record the amount by cutting notches in a tally-stick and after the notches were made the tally-stick was split down the middle, the creditor retaining one half and the debtor the other. Hence such expressions as "Our figures don't tally."

The British Government kept its records of transactions by the tally system up to 1543. In 1834, when the accumulation of old tally-sticks had grown beyond reasonable limits, it was decided to burn them. The stove in which they were being burned was over-stoked and set fire to the wainscoting, and soon the whole room was ablaze. So great a fire ensued that both Houses of Parliament were burned down.

THE FIRST CALCULATING MACHINES

The first calculating machine was devised by Blaise Pascal, a Frenchman, in 1642. He was only nineteen

BLAISE PASCAL
After a contemporary drawing

when he built an adding machine to help his father, who was a Government accountant. Its principle was the now-familiar device of a series of geared wheels each having ten teeth, one for each unit from 0 to 9. At each complete turn of any wheel the next higher one was turned through a tenth of a revolution. When we see the speedometer of a car, the cyclometer on a bicycle, or the homely gas or electric meter, we should remember that they owe their existence to Pascal's adding machine.

Many other men tried their hands at inventing machines which multiplied, as well as adding and subtracting, and could deal with sums of money as well as numbers.

Nowadays many firms market machines which enable costing calculations which formerly took a long time—particularly when errors had to be detected and corrected—to be done in a matter of seconds.

MODERN CALCULATING MACHINES

Most commercial calculations are, basically, those of addition and subtraction; for multiplication is a convenient shortening of repeated addition, division of repeated subtraction. So the fundamental principle of calculating machines is accurate mechanical addition and subtraction.

The machines can be used for all kinds of commercial work. The totalling of accounting books and documents, clock cards, job cost cards, counterfoils, and invoices are examples of addition operations for which they are invaluable. The multiplication and division work which they can do includes the extension and checking of sales and purchases  invoices; wages calculations, including bonus, overtime and piece-work rates, and P.A.Y.E.; profit-and-loss calculations; and simple and compound interest.

We shall not enter here into the mechanics and manner of use of the various machines; the companies which make and market them generally maintain schools where operators are trained in a short time. The Sumlock standard sterling model adding-calculating machine, shown in the illustration at p. 291, is one of the products of London Computator, Ltd. A feature of this machine is that it 'locks' itself automatically if an error is made.

In addition to the usual range of machines Burroughs Adding Machine, Ltd, produce a typewriter-accounting and calculating machine. This machine combines the automatic features of the electrically operated adding-subtracting book-keeping machine, a full standard keyboard typewriter, and a calculating mechanism, which enables not only totals and balances, but also multiplication products to be computed and printed by depression of a single key.

With a modern accounting machine it is claimed that such an operation as 5 tons 12 cwt. 3 qr. at £4 7s. 8d. a ton can be performed in about four seconds; the calculation of what percentage one sum of money is of another takes about six seconds; while the addition of an eighteen-row money sum including hundreds of £ as well as shillings and pence is a matter of fourteen or fifteen seconds.

Machines have also been invented for solving equations and for doing mechanically quite complicated operations in the higher mathematics. But enough has been said to indicate the vast forward strides that are being made in this department of computation.

Yet with all these swift advances in time and labour saving the necessity still arises for alert, competent operators; and the quickness of mind which training in elementary mathematics should produce is still basic even in mechanical computation.

PART VI

PEEPS OVER THE FENCE

CHAPTER XXII

SOME MIXED TOPICS

(1) ADDITION RULES

MOST people are so accustomed to the usual rules of arithmetical addition that it may come as a surprise to them to know that there are other ways of adding quantities. And when we refer to arithmetic algebra is included; for finding such a sum as $2a + 3b - 4a + 9c + 7a - 3b$ is, after all, not very different from the addition of money or weights and measures, except that there are no 'carrying' figures such as arise from the fact that there are, for instance, twelve pence in a shilling or sixteen ounces in a pound.

We shall in this section touch briefly upon some other addition rules. In some cases we may be able to demonstrate their validity; in others we shall just state the facts, the proof being a matter of more advanced mathematics.

LOGARITHMIC ADDITION

We saw at p. 276 that
$$\log x + \log y = \log xy$$
(and *not* that $\log x + \log y = \log (x + y)$);

also, that $\quad \log x - \log y = \log \dfrac{x}{y}$

(and *not* that $\log x - \log y = \log (x - y)$).

293

As a matter of interest:

$$\log (x + y) = \log x + \log y \text{ only when}$$
$$\log xy = \log (x + y).$$

We have, then, $\qquad xy = x + y;$

that is, $\qquad\qquad xy - y = x;$

$$y(x - 1) = x;$$

$$y = \frac{x}{x - 1}.$$

Can you deduce a similar expression for the special case when

$$\log (x - y) = \log x - \log y?$$

VECTOR ADDITION

If I set out on a three-mile walk, and the spirit moves me to make a detour and thus add two miles to my walk, I shall probably very proudly boast on my return that I have walked five miles.

If I walk three miles in a northerly direction, and then turn east and continue for another two miles, I shall cer-

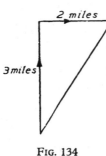

FIG. 134

tainly have done a five-mile walk; but I shall still be very far from home, and my final position will depend only partly on the fact that I have walked five miles. To get my final position and my direct distance and direction from home I shall want something of the nature of the scale drawing of Fig. 134. Of course, I can calculate in this particular case my distance from home by Pythagoras's theorem, and my direction from the trigonometry of the right-angled triangle.

These two instances are given as illustrations of *scalar* and *vector* quantities.

Scalar quantities have magnitude only; vector quantities have magnitude and direction. We shall give more examples of them later.

The addition of scalar quantities is a matter of simple arithmetic; the addition of vectors is done by one of two laws—the *triangle law* and the *parallelogram law*.

Adding by the triangle law is done as in Fig. 135. The quantities—we shall call them p and q—which are given in magnitude and direction are drawn to scale and placed end to end (Fig. 135), their directions being indicated by single barbed arrows. The arrows

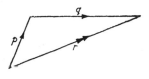

FIG. 135

must 'flow' in the same direction round the sides of the triangle. The sum (r)—or *resultant*, as it is called—is given in magnitude and direction by the third side of the triangle, and is usually indicated by a double-barbed arrow.

Adding by the parallelogram law is done by placing the scaled lines representing the quantities next to each other as in Fig. 136, with the directional arrows in this case pointing outward from the point from which the lines are drawn. The parallel-

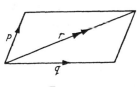

FIG. 136

ogram of which the lines representing p and q form two adjacent sides is then completed. The diagonal of this parallelogram through the point where the lines representing p and q meet represents in magnitude and direction the sum (r), or resultant, of the vectors. Addition by the triangle law or the parallelogram law will give exactly the same result.

In this form of addition the drawing of the lines to scale and the subsequent measuring of the resultant takes the place of the processes used when the ordinary methods of arithmetical addition are employed. It may, in a manner of speaking, be compared with the addition process when two rules sliding on each other are used.

Other examples of vector addition are the resultants of a load being hauled by means of two ropes not parallel to each other; a boat or ship being propelled in one direction, the tide or current running in a different direction; an aeroplane headed in one direction, in a wind blowing in a different direction.

Notes. (1) The principle of the triangle law may be extended to any number of vectors, provided they all act at or pass through a single point. Their sum, or resultant effect, may be obtained by laying the scaled lines representing them end to end in their proper directions; the order is immaterial so long as their directions 'flow.' The sum, or resultant, will then be represented in magnitude and direction (but not, of course, in position) by the vector represented by the line connecting the starting-point with the final point reached. In particular, if the final point coincides with the starting-point the sum is zero.

(2) Regarding a subtraction as an addition with the sign of the subtracted quantity reversed, we see that one vector may be subtracted from another by reversing its direction (or *sense*, to use the proper technical term) and adding the vector so reversed.

ADDITION RULES IN TRIGONOMETRY

Each question here may be put in three forms; *e.g.*, using sines:

(i) What is $\sin x + \sin y$?

(ii) What is $\sin (x + y)$?

(iii) Are they equal to each other?

Similar questions may be proposed for the cos-ratio and the tan-ratio.

We can test the truth or otherwise of the third form of question by a simple substitution.

Suppose $x = 30°, y = 30°$.
Then, since sin $30° = \cdot5$ (calculation or tables),
$\quad$ sin x + sin y = $\cdot5$ + $\cdot5$ = $1\cdot0$.
But $\quad x + y$ now = $60°$;
$\quad$ and sin $(x + y)$ = sin $60°$ = $\cdot8660$.
So sin x + sin y is not equal to sin $(x + y)$.
Similarly, cos x + cos y is not equal to cos $(x + y)$.
And tan $(x + y)$ is not equal to tan x + tan y.

In point of fact, the simple addition rules in trigonometry are given by:

$$\sin(x + y) = \sin x \cos y + \cos x \sin y;$$

$$\sin x + \sin y = 2 \sin \frac{x + y}{2} \cos \frac{x - y}{2};$$

$$\cos(x + y) = \cos x \cos y - \sin x \sin y,$$

$$\cos x + \cos y = 2 \cos \frac{x + y}{2} \cos \frac{x - y}{2}.$$

$$\tan(x + y) = \frac{\tan x + \tan y}{1 - \tan x \tan y}.$$

There are several forms for tan x + tan y; not one of them is very elegant.

(II) STATISTICS

THE NEED FOR THE STUDY OF STATISTICS

We may almost call statistics one of the 'new' subjects in mathematics; or perhaps it would be more nearly true to say that the modern approach to the science of statistics is a new and rapidly growing subject.

There are large classes of men to whom the study of statistics is of great value. These include not only actuaries and accountants, but also social workers, politicians (national and local), and business men. For, unless figures are properly handled, conclusions drawn from them are likely to be seriously misleading.

It is a commonplace that figures may be made to prove anything. For example, in October 1945 103 people were killed on the road; in October 1944 the figures were 95. On the face of it there was an *increase* of eight. But when it was pointed out that there were three times as many vehicles on the roads in October 1945 as there were in October 1944 it was realized that an apparent increase was really a comparative decrease.

AVERAGES

It has already been shown in Chapter VIII how the study of certain facts is simplified when they are collected in tabular form and then presented in pictorial form as graphs. This was our real introduction to the study of statistics.

We are also familiar with averages. The average of a group of quantities is obtained by finding their sum and dividing this sum by the number of quantities in the group. Averages are the first step in the further study of statistics. Knowledge of his average sales is often of more importance to a business man than that of his sales on particular days; the average output of a workman or —what is not the same thing—the output of the average workman, is essential in fixing the time taken for a job or in 'costing' a process.

But averages too may be misleading. If the greatest (or smallest) of a number of quantities is far removed from the general run, the calculated average will not be a true indication of what Mr Everyman would regard as the average. It is for this reason that before striking an average some people leave out any facts or figures which deviate greatly from the general run.

By the way, the technical term for an average of a number of quantities is the *mean*, or the *arithmetic mean*, of the quantities.

THE MEDIAN

If a set of numerical facts is arranged in order of size starting from the greatest, or the least, then that item which is such that the same number of items lies above it as below it is said to be the *median*.

Thus, if nine men are weighed and their weights are 140, 152, 138, 145, 164, 158, 142, 147, 160 lb., to find the median weight we arrange them as follows:

164, 160, 158, 152, | 147 |, 145, 142, 140, 138.

The median weight is 147 lb., for there are four weights less than it and four weights greater than it.

If there is an even number of items the median is taken as the mean (or average) of the two middle quantities.

The advantage of the median over the mean is that it is easier to find, particularly when the number of items is large.

A teacher who wishes to compare the standard of attainment of several classes who have been set the same test can do so better by finding the median than by examining the lists of marks for each class. Also, the value of the median is not affected by extreme cases like a very high mark or a very low mark, as the average of the marks would be.

EXERCISE 39

(a) Find the average, (b) find the median, of the following sets:

1. 23, 29, 27, 35, 34, 25, 31, 28, 24, 30, 22.
2. 126, 148, 133, 153, 138, 129, 140, 136, 130.
3. 416, 433, 429, 438, 420, 441, 419, 436.

FREQUENCY

Suppose that a hundred men are picked at random from a crowd and that their heights to the nearest half-inch are measured and tabulated. It is exceedingly probable that many of them will have the same height.

Or again, suppose that fifty pods of peas freshly picked from a row are opened and the numbers of peas in each pod are tabulated. Many of the pods will contain equal numbers of peas; a few will be of the kind that gladden the gardener's heart, others . . . !

If tables are arranged to show how many times each particular rating occurs (that is, how many of the men are 5 ft. 8 in. high, or how many pods contain six peas, and so on) they are called *frequency tables*. A frequency table shows the *distribution* of the *ratings* or items.

A frequency table may be useful in forecasting, for instance, how many of each size of clothing will be required by an outfitter for next season's stock.

Suppose that fourteen pods of peas contained 8, 6, 7, 7, 5, 9, 7, 6, 4, 7, 8, 5, 6, and 7 peas. These would be arranged in a frequency table horizontally as follows:

No. of peas · ·	4	5	6	7	8	9
Frequency · ·	1	2	3	5	2	1

The frequency table may also be arranged in vertical columns.

The item which occurs most frequently is called the *mode* or the *modal* item. In the case of the fourteen pods of peas the modal number of peas is seven, with frequency five.

In making a frequency distribution for a large number of items it may be well to group the data into a smaller number of equal intervals; *e.g.*, if the number of eggs laid by a group of hens each day for a month was 25, 28, 27, 32, 35, 34, 38, 26, . . . and so on, it would be better, in making a frequency table, to group the numbers by 3's or 4's, thus: 25–27, 28–30, . . . or 25–28, 29–32, ...

EXERCISE 40

1. Arrange the following sets in frequency tables, indicating in some special manner the mode in each case:

(a) Temperatures: 66°, 74°, 70°, 72°, 74°, 70°, 74°, 72°, 70°, 68°, 76°, 68°, 72°, 70°.

(b) Occurrences of dice: 6, 4, 5, 5, 4, 2, 1, 3, 6, 4, 3, 2, 5, 4, 3, 4.

2. The following marks occurred in a test. Arrange them in a frequency table, grouping the marks in fives from 30 to 79:

47, 62, 75, 34, 52, 50, 48, 64, 79, 31, 62, 49, 54, 41, 59, 52, 48, 72, 44, 53.

What is the modal group?

HISTOGRAMS

Frequency tables may be depicted graphically by means of upright bar graphs, the bars being made to touch each other. Frequency bar graphs are called *histograms* (*histos*, Greek for 'mast'). Alternatively, a *frequency curve* may be drawn. In each case the cross-axis is used for the various items or ratings, the upright axis for the frequency. A histogram may be converted into a frequency curve by connecting the middle points of the tops of the bars. Clearly the longest bar, or the greatest height in the case of a curve, will be the mode.

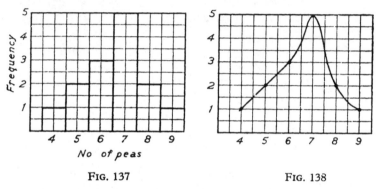

FIG. 137   FIG. 138

Figs. 137 and 138 show a histogram and a frequency curve for the trial with the fourteen peas at p. 300.

THE CENTRAL TENDENCY

When statistical data are arranged in frequency tables or depicted by means of histograms or frequency curves it will often be noticed that there is a tendency for them to be grouped round a centre. Most of the ratings appear to become concentrated towards the middle of the distribution, which tapers at each end because there are usually very few high ratings and very few low ratings. After all, in a large group of persons the tendency will be for there to be very few near-giants and very few near-dwarfs.

It must not be supposed that this shape is character-

302

istic of all histograms and frequency curves, but it does result from fields in which the law of chance seems to operate. Examples that may be given are the findings of biological and economic experiments and investigations, heights or weights of a large sample of people, and marks in a test or in intelligence tests. Natural phenomena seem to vary continuously and according to this form of distribution.

This appears to be a fundamental law of nature. If we could

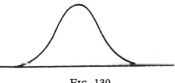

FIG. 139

test the intelligence of all the people of Great Britain of a certain age the frequency curve for the result of the tests should be a smooth curve something like that shown in Fig. 139.

The present writer has found from many years' experience of marking School Certificate papers of one of our great examining boards that, though the marks gained by the pupils in any one school may vary greatly, the marks for five hundred pupils taken from many schools tend to arrange themselves according to this law.

This curve is called the *curve of normal distribution.* In a theoretically perfect normal distribution the median, the mean, and the mode are identical.

The curve of normal distribution has symmetry about the mode, and it tapers off equally at each end. It is the curve which would be produced if in collecting data only the law of chance, free from any of the outside influences which enter into human situations, operated.

RANDOM SAMPLING

Of course, it would not be possible for any one person or number of persons to test the intelligence of everybody in the world, or even in a single country, or to count the number of peas in every pod in every row in a field.

Fortunately it is not necessary for this to be done. If we can test a random sample—that is, a sample selected purely by chance—of a group we can get a picture of what the whole is likely to be. If we were to take, say, one hundred pods of peas at random from a row and count the number of peas in each pod we should find that the mean, the mode, and the distribution would be almost the same as for any other hundred pods selected at random, and so, by inference, the same as for all the pods in all rows in the field.

The important thing is that the sample should be selected at random. Hence, for instance, the care taken by the various institutes which test public opinion on its reaction to current problems. If the sample taken is not a random cross-section of the people whose opinions are desired the conclusions deduced therefrom are valueless.

More and more use of random sampling is being made by manufacturers and business men. Results deduced from random samples—and it must again be stressed that this means samples selected by chance—are taken as true criteria for the whole group, be it in tests of a qualitative or quantitative nature or of reactions to an advertisement campaign.

The methods of random sampling are still a matter of scientific investigation, the essential point being the number of samples necessary in order that errors may be confined within reasonable limits.

EXERCISE 41

1. Heights were taken of 840 men chosen at random from a football crowd in Great Britain. Depict the following data first by a histogram, then by a frequency curve:

Height in inches	Under 62	63–64	65–66	67–68	69–70	71–72	Over 73
Frequency	30	100	210	250	170	60	20

2. During one of the War years an analysis of the ages of drivers involved in road-vehicle collisions was made. The facts were:

Under 21 years of age . .	8%
21–30	26%
31–40	37%
41–50	20%
Over 50	9%
	100%

Show these figures first by a histogram, then by a frequency curve.

TAILPIECE

Only the elements of the theory of statistics have been given here. A serious study of statistics involves quite advanced mathematics. But the reader who has grasped the significance of the curve of normal distribution, and who has realized the advantages of the median over the mean, has at any rate taken a good long peep over this particular fence!

(III) THE INVESTIGATION OF FUNCTIONAL RELATIONSHIPS, OR MATHEMATICAL EXPLORATION

We saw at p. 256 that mathematical data may be presented in three different ways:

(*a*) statistical data may be collected from practical experiments by, for example, the engineer and the scientist, and exhibited in tabular form; or

(*b*) they may be illustrated conveniently by means of a curve; and

(*c*) theoretical considerations may lead to the formulating of a functional relationship.

Let us now take our discussion a stage farther. To begin with, we have seen at pp. 245, 246, that if a set of points is found, when plotted, to lie on a straight line there is a functional relationship of the first degree between their co-ordinates. In the same way, if it can be shown that when, say, log y is plotted against log x for given numerical values of x and y the points so plotted lie on a straight line, there is a functional relationship of the form

$$y = kx^n,$$

where k and n can be evaluated for the particular case under consideration. These are examples of obtaining the relationship from (*b*).

The determination of a functional relationship directly from (*a*) raises more difficulties. Robert Boyle (1627–91) did it. He noticed that if some gas was enclosed in such a way that its volume and pressure could be varied without altering its temperature, the product of the volume (*v*) and the pressure (*p*) was approximately constant; in symbols:

$$pv = \text{constant} = k \text{ (say).}$$

This conclusion was deduced from actual experimental results, and became known as Boyle's Law. Later it was found that the law was true only under special conditions, and Van der Waals (1837–1923) showed by theoretical considerations that a better approximation to the behaviour of the gas was given by the relation:

$$(p + \frac{a}{v^2})\,(v - b) = \text{constant},$$

where a and b are numerical values always having the same value for each particular gas. Even this relation has been shown to be only a further approximation.

The investigation of such problems as this has sometimes given rise to entirely new mathematical functions and literature in connexion with them. Sometimes a lucky guess has led the way to the answer; at other times careful theorizing allied with experimental testing has led to relations consistent with the given facts. Newton's Law of Universal Gravitation and Einstein's Theory of Relativity are cases in point.

To take another example: a person who has done only elementary mathematics would be hard put to find any connexion in the sequence of numbers

$$6 \quad 20 \quad 40 \quad 66 \ \ldots$$

which would enable him to write down subsequent numbers in the sequence and ultimately its sum to any given number of terms.

A mathematician would probably start by trying a 'difference' method. He writes down a table in which each new line is formed by subtracting each number in the next line above it from the next number in that line. This gives

$$
\begin{array}{ccccccc}
6 & & 20 & & 40 & & 66 \\
& 14 & & 20 & & 26 & \\
& & 6 & & 6 & & \\
& & & 0 & & &
\end{array}
$$

From the fact that the third line of differences is 0 he can proceed in either of two ways.

(*a*) He can continue the table by building it up, as it were, backward, giving

$$6 \quad 20 \quad 40 \quad 66 \quad (98)$$
$$14 \quad 20 \quad 26 \quad (32)$$
$$6 \quad 6 \quad (6)$$
$$0 \quad (0)$$

The next figures are shown here in parentheses. Do you see how they are obtained? And can you find the next two numbers in the sequence?

Or, (*b*) by theoretical considerations built up on work which has been developed by mathematicians in the past, he can say that the *n*th term of the series is of the form

$$an^2 + bn + c,$$

where *a*, *b*, and *c* are numerical 'constants' which can be easily determined from the first three terms. Of course, these 'constants' change with each separate sequence of terms. Further, the sum of *n* terms of the sequence is known to be of the form

$$pn^3 + qn^2 + rn + s$$

where *p*, *q*, *r*, and *s* are numerical 'constants' which can be determined from the first four terms of the sequence. It should be added that the degree in *n* of the sum of *n* terms is the number of the difference line which is first zero; and the degree in *n* of the *n*th term is one lower than this.

This, of course, is only *one* of the tools at the disposal of the mathematician. It solves certain problems. But even with the advance in mathematical techniques he is still only on the edge of the field; he too is just peeping over the fence.

CHAPTER XXIII

THE CALCULUS

THE APPROACH

LET us consider $y = x^2$. y is a function of x; to every value of x there is a corresponding value of y. If, when x is increased by a small quantity h, y becomes $y + k$, then $y + k = (x + h)^2 = x^2 + 2xh + h^2$ (see p. 214).

But $y = x^2$.

So, subtracting,

$$k = 2xh + h^2,$$

or $$\frac{k}{h} = 2x + h,$$ on dividing through by h.

If now we take $x = 1$, so that $y = 1$, we have $\frac{k}{h} = 2 + h$

If $h = \frac{1}{10} = \cdot 1$, then $\frac{k}{h} = 2 + \cdot 1 = 2 \cdot 1$;

if $h = \frac{1}{100} = \cdot 01$, then $\frac{k}{h} = 2 \cdot 01$;

if $h = \frac{1}{1000} = \cdot 001$, then $\frac{k}{h} = 2 \cdot 001 \cdot$

and so on.

It is clear that as we make h smaller and smaller, the value of $\frac{k}{h}$ will differ from 2 by an amount which also gets smaller and smaller. We say that the limit of the value of $\frac{k}{h}$ as h tends to 0 (written $h \to 0$) is 2, and we write it

$$\operatorname*{Lt}_{h \to 0} \frac{k}{h} = 2.$$

In the meantime, let us see what has been happening to the value of k.

For $x = 1$, $k = 2h + h^2$;
and for $h = \cdot 1$, $k = \cdot 2 + \cdot 01 = \cdot 21$;
and for $h = \cdot 01$, $k = \cdot 02 + \cdot 0001 = \cdot 0201$;
and for $h = \cdot 001$, $k = \cdot 002 + \cdot 000001 = \cdot 002001$.

Thus, as h tends to zero, so does k, but not in exactly the same way.

We have, then, two quantities k and h which both tend to zero and yet whose ratio has a definite, finite value.

This discussion may at first seem rather academic. Can we find its counterpart in real life? We can, and will.

What is meant by the speed at any moment of, say, a car? It is the rate at which the car is travelling. That such a speed exists at every moment of travel the speedometer, which is an arrangement of geared wheels transmitting the rotation of a wheel of the car to an indicator, tells us; but how can we arrive at this speed by calculation? When we talk about a speed of 30 m.p.h. our unit of time is an hour. But how does this help us to fix a speed on a run at some definite time, which for convenience of reference we shall say is 4 P.M.?

If we divide the distance travelled between 3.55 P.M. and 4.5 P.M. by the time—that is, 10 min., or $\frac{1}{6}$ hour—we have the average speed over a period of 10 minutes. But any motorist knows that in 10 minutes the actual speed of a car may vary considerably. Even if we obtain the average speed from 3.59$\frac{1}{2}$ P.M. to 4.00$\frac{1}{2}$ P.M. there is still a possibility of this variation.

By now the reader must realize what we are getting at. The actual speed at 4 P.M. is the limit of the value of the expression $\dfrac{\text{distance}}{\text{time}}$

when the time, and consequently the distance, tends to zero.

THE CALCULUS NOTATION

Let us return to our discussion of the function $y = x^2$. If instead of h and k (at p. 309) we write δx and δy (the δ is the Greek small letter d) we can say (see p. 309):

$$\frac{\delta y}{\delta x} = 2x + \delta x.$$

If now we make δx tend to 0—we write this as $\delta x \to 0$ —our discussion enables us to say:

$$\underset{\delta x \to 0}{\mathrm{Lt}}\ \frac{\delta y}{\delta x} = 2x.$$

This is rather cumbersome to write, and so a special notation was devised. It is:

$$\frac{dy}{dx} = 2x.$$

Note. δy and δx are separate small quantities; $\frac{dy}{dx}$ is just a convenient way of writing down the limit when $\delta x \to 0$ of $\frac{\delta y}{\delta x}$. It does not mean $dy \div dx$. dy and dx here do not exist independently. (Possibly $\frac{dy}{dx}$ might be written as $\frac{d}{dx}(y)$ to indicate that a certain operation has been performed on y.) We read it: "dee-y by dee-x" (and never "dee-y divided by dee-x").

This fact—that if $y = x^2$ then $\frac{dy}{dx} = 2x$—is a special case of a more general result that can be proved:

$$\text{if } y = x^n,$$
$$\text{then } \frac{dy}{dx} = nx^{n-1}.$$

To write down $\frac{dy}{dx}$ here all we have to do is to bring the index of x (that is, n) down as a coefficient and subtract 1 from the index of x.

If y is a constant—that is, quite independent of x—there can be no change in the value of y whatever happens to x. So:

$$\text{if } y = c, \text{ then } \frac{dy}{dx} = 0.$$

Also, if $y = ax^n$ it can be proved that

$$\frac{dy}{dx} = nax^{n-1}.$$

For example, if $y = 8x^5$, then $\frac{dy}{dx} = 40x^4$.

And if $y = 4x^3 - 7x^2 + 9x - 5$, then $\frac{dy}{dx} = 12x^2 - 14x + 9$; the same process is carried out separately on each power of x, the connecting signs remaining the same.

Calculating, or writing down, the value of $\frac{dy}{dx}$ when y is a given function of x is called *differentiating the function with respect to x*; and the study of this and allied matters is called the *differential calculus*. The process of differentiating is also called *finding the differential coefficient of y with respect to x*.

EXERCISE 42

1. Find $\frac{dy}{dx}$ when $y = x^3$; $y = 2x$; $y = 5x + 3$; $y = 7x - 4$; $y = 3x^2$; $y = 10x^7$; $y = 5x^2 - 4x$; $y = 7x^2 + 8x - 9$; $y = 3x^3 - 2x^2 + x - 1$.

Find $\frac{ds}{dt}$ when $s = 3 + 4t$; $s = 8 - 6t$; $s = 16t^2$; $s = 10t + 16t^2$; $s = 80t - 16t^2$.

REVERSING THE PROCESS

We have seen that if $y = x^2$, then $\dfrac{dy}{dx} = 2x$. We can reverse the process and say:

if $\dfrac{dy}{dx} = 2x$, then $y = x^2$—or, more accurately,

$$y = x^2 + c,$$

where c is any constant; for differentiating x^2 and $x^2 + c$ with respect to x gives us the same result.

This reversed process is called *integrating*, and the study of it and its consequences is called the *integral calculus*. x^2, or $x^2 + c$, is called the *integral* of $2x$.

There is also a special notation for integration.

If $\dfrac{dy}{dx} = 2x$ we write down the integral as $y = \int 2x dx$, the $\int \ldots dx$ indicating that the expression under the $\int$ sign is to be integrated with respect to x. We read it: "integral . . . dee-x."

In general, if $y = x^n$,

$$\int y dx = \int x^n dx = \frac{x^{n+1}}{n+1} + c.$$

Here to integrate x^n we increase the index of x by 1 and divide by the index so increased.

For example,

$$\int (3 - 4x + 2x^2 - x^3) dx = 3x - \frac{4x^2}{2} + \frac{2x^3}{3} - \frac{x^4}{4} + c$$

$$= 3x - 2x^2 + \frac{2}{3}x^3 - \frac{x^4}{4} + c.$$

EXERCISE 43

1. Write down the integrals with respect to x of:
   x, x^3, $5x^2$, $3x + 4$, $5 - 2x$, $2x^2 + 2x + 1$, $3 - 4x - 3x^2 + 4x^3$.

2. Write down the integrals with respect to t of:
   $2t$, $3t + 2$, $4 - t^3$, $3t^2 + 2t - 7$, $4 - 8t + 6t^2$.

GEOMETRICAL ILLUSTRATION OF DIFFERENTIATION

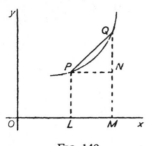

FIG. 140

Suppose that the relationship between y and x is depicted graphically by the graph of Fig. 140.

Let P and Q be two points on the graph whose co-ordinates (p. 245) are (x, y) and $(x + \delta x, \quad y + \delta y)$ respectively. Then if PL, QM are drawn perpendicular to the axis of x, and PN is drawn perpendicular to QM, we have:

$$\frac{QN}{PN} = \frac{QM - NM}{PN} = \frac{QM - PL}{LM} \quad \text{(for the opposite sides of a rectangle are equal)}$$

$$= \frac{(y + \delta y) - y}{\delta x} = \frac{\delta y}{\delta x}.$$

So if $Q\hat{P}N = \theta$—that is, θ is the angle which QP makes with the axis of x—then $\tan \theta = \dfrac{\delta y}{\delta x}$ (see p. 193).

What can we say in this case when $\delta x \to 0$? We see that for this to happen Q must approach nearer and nearer to P. In the limit, $\dfrac{\delta y}{\delta x} \to \dfrac{dy}{dx}$, Q and P coalesce, and QP touches the curve at P.

So $\dfrac{dy}{dx}$ is the tangent of the angle made by a straight line which touches the curve which represents graphically the relationship between y and x. Of course, to obtain a numerical result we must know the numerical value of the co-ordinates of P.

Thus if the curve represents $y = x^2$ we know that

314

$\frac{dy}{dx} = 2x$, and the tangent of the angle made with the axis of x by the straight line which touches this curve at the point $(1, 1)$ (check that this point lies on the curve) is given by $\tan \theta = 2$, which from the tables gives $\theta = 63\frac{1}{2}°$ (about).

GEOMETRICAL ILLUSTRATION OF INTEGRATION

We start with P and Q as before. If we also draw QR perpendicular to LP produced (Fig. 141) then the area $PLMN$ is less than the area $PLMQ$, where $PLMQ$ means the area bounded by the curve between P and Q and by PL, LM, and MQ.

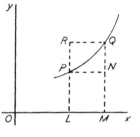

FIG. 141

That is, $y \times \delta x$ is less than the area $PLMQ$.

Similarly, the area $PLMQ$ is less than the area $RLMQ$; that is, the area $PLMQ$ is less than $(y + \delta y)\delta x$.

So the area $PLMQ$ lies between $y\delta x$ and $y\delta x + \delta y \delta x$. But when $\delta x \to 0$, $\delta y \delta x$ will be very small indeed, and so the small area—or element of area, as it is called—$PLMQ \to y\delta x$. Suppose we call this area δA (A for area). Then

$$\delta A \to y\delta x.$$

Now any area bounded by the curve and the x-axis and the perpendiculars drawn to the x-axis from two stated points on the curve will consist of the sum of small areas like δA.

So the sum of the δA's = the sum of the $y\delta x$'s;

that is, the whole A = the sum of the $y\delta x$'s.

315

But, using the notation first of the differential calculus, then of the integral calculus, we have, since $\delta A \to y \delta x$,

$$\frac{dA}{dx} = y;$$

and so $A = \int y dx$.

We see thus that $\int y dx$ = the sum of the $y \delta x$'s; that is, that integration, as well as being the reverse process of differentiation, may be regarded as a summation—in particular, that of the area 'under' a curve.

TWO INTERESTING APPLICATIONS

If the speed of a car is plotted against the time measured from the beginning of a trip (or, in fact, from any definite time) the area 'under' the speed-time curve between two stated times will give the distance travelled in that time, however much the speed may have varied in that time interval. The computation of the area may be done in any manner—for instance, by counting squares.

Moreover, the slope of the curve at any point will indicate whether the car is being accelerated or retarded, or is travelling at steady speed. For, as we have seen, the tangent of the angle made with the t-axis (not the x-axis in this case) by a straight line which touches the curve will be equal to $\frac{dv}{dt}$ where v is the velocity, or speed; and this is the limit when $t \to 0$, of $\frac{\delta v}{\delta t}$, which is the limit of the increase of velocity divided by the time in which it has taken place.

If $\frac{dv}{dt}$ is positive there is acceleration; if $\frac{dv}{dt} = 0$ there is no change in velocity; if $\frac{dv}{dt}$ is negative there is re-tardation.

316

In general, we can regard the differential coefficient as a measure of the 'rate of growth' of a function. The greater the value of the differential coefficient, the greater is the rate of growth when that rate is positive. If, on the other hand, the differential coefficient is negative the 'rate of growth' is negative; that is, there is decrease, not increase.

HOW THE CALCULUS DEVELOPED

Of course, in this brief introduction it is impossible to do more than indicate the meaning of the calculus symbolism. Little did Sir Isaac Newton, to whom British mathematicians ascribe the honour of inventing the calculus, think what a powerful instrument he was forging for the promotion of future advances in all scientific subjects.

It should be mentioned, however, that Continental mathematicians maintain that the primary credit should be given to a German named Gottfried Leibnitz. Newton began developing his ideas of the calculus in 1665. In 1671 he wrote his treatise on fluxions, the forerunner of the calculus,

ISAAC NEWTON

but it was not published until 1736, nine years after his death. According to his own account, Leibnitz began his work on the calculus about 1674, but his work, though completed after Newton's, was published before Newton's was. And there the controversy rests. Possibly Leibnitz did more to establish the present symbolism; on the other hand, Newton was the pioneer, and he laid the foundations of the mathematical theory with more rigour than Leibnitz did.

This is just a peep at the calculus; it is a vast subject with numerous ramifications. The reader whose interest has been aroused is advised to get hold of a copy of *Calculus Made Easy*, by Sylvanus P. Thompson, F.R.S. (Macmillan). It is a most valuable introduction to the subject, and has the additional merit of being extremely readable.

CONCLUSION

Our course is completed. It is sketchy in many respects; it is necessarily dull in parts. But the author would here repeat some words which he used in the preface: no one will expect to find a book about mathematics which can be read quite like a novel or a history book.

The reader, on his part, will want to know, "How do I stand now as regards some definite educational standard in mathematics, such as the School Leaving Certificate, if I have normal intelligence and have studied the book faithfully?" The answer, roughly, is: in arithmetic the course has been almost covered, certainly the principles; in algebra and trigonometry, judged by present-day tendencies, much of it has been covered; in geometry much more would have to be done, particularly in demonstrative geometry. The glimpses given of Statistics, the Calculus, mechanical aids, etc., in the last three chapters are of general interest and utility, but they are not included in the ordinary syllabuses at this stage. It is our belief that historical sketches add much to the enjoyment of any subject, and some day more may be made of them in the teaching in our schools.

LOGARITHMS

	0	1	2	3	4	5	6	7	8	9	1 2 3	4 5 6	7 8 9
10	0000	0043	0086	0128	0170	0212	0253	0294	0334	0374	4 8 12	17 21 25	29 33 37
11	0414	0453	0492	0531	0569	0607	0645	0682	0719	0755	4 8 11	15 19 23	26 30 34
12	0792	0828	0864	0899	0934	0969	1004	1038	1072	1106	3 7 10	14 17 21	24 28 31
13	1139	1173	1206	1239	1271	1303	1335	1367	1399	1430	3 6 10	13 16 19	23 26 29
14	1461	1492	1523	1553	1584	1614	1644	1673	1703	1732	3 6 9	12 15 18	21 24 27
15	1761	1790	1818	1847	1875	1903	1931	1959	1987	2014	3 6 8	11 14 17	20 22 25
16	2041	2068	2095	2122	2148	2175	2201	2227	2253	2279	3 5 8	11 13 16	18 21 24
17	2304	2330	2355	2380	2405	2430	2455	2480	2504	2529	2 5 7	10 12 15	17 20 22
18	2553	2577	2601	2625	2648	2672	2695	2718	2742	2765	2 5 7	9 12 14	16 19 21
19	2788	2810	2833	2856	2878	2900	2923	2945	2967	2989	2 4 7	9 11 13	16 18 20
20	3010	3032	3054	3075	3096	3118	3139	3160	3181	3201	2 4 6	8 11 13	15 17 19
21	3222	3243	3263	3284	3304	3324	3345	3365	3385	3404	2 4 6	8 10 12	14 16 18
22	3424	3444	3464	3483	3502	3522	3541	3560	3579	3598	2 4 6	8 10 12	14 15 17
23	3617	3636	3655	3674	3692	3711	3729	3747	3766	3784	2 4 6	7 9 11	13 15 17
24	3802	3820	3838	3856	3874	3892	3909	3927	3945	3962	2 4 5	7 9 11	12 14 16
25	3979	3997	4014	4031	4048	4065	4082	4099	4116	4133	2 3 5	7 9 10	12 14 15
26	4150	4166	4183	4200	4216	4232	4249	4265	4281	4298	2 3 5	7 8 10	11 13 15
27	4314	4330	4346	4362	4378	4393	4409	4425	4440	4456	2 3 5	6 8 9	11 13 14
28	4472	4487	4502	4518	4533	4548	4564	4579	4594	4609	2 3 5	6 8 9	11 12 14
29	4624	4639	4654	4669	4683	4698	4713	4728	4742	4757	1 3 4	6 7 9	10 12 13
30	4771	4786	4800	4814	4829	4843	4857	4871	4886	4900	1 3 4	6 7 9	10 11 13
31	4914	4928	4942	4955	4969	4983	4997	5011	5024	5038	1 3 4	6 7 8	10 11 12
32	5051	5065	5079	5092	5105	5119	5132	5145	5159	5172	1 3 4	5 7 8	9 11 12
33	5185	5198	5211	5224	5237	5250	5263	5276	5289	5302	1 3 4	5 6 8	9 10 12
34	5315	5328	5340	5353	5366	5378	5391	5403	5416	5428	1 3 4	5 6 8	9 10 11
35	5441	5453	5465	5478	5490	5502	5514	5527	5539	5551	1 2 4	5 6 7	9 10 11
36	5563	5575	5587	5599	5611	5623	5635	5647	5658	5670	1 2 4	5 6 7	8 10 11
37	5682	5694	5705	5717	5729	5740	5752	5763	5775	5786	1 2 3	5 6 7	8 9 10
38	5798	5809	5821	5832	5843	5855	5866	5877	5888	5899	1 2 3	5 6 7	8 9 10
39	5911	5922	5933	5944	5955	5966	5977	5988	5999	6010	1 2 3	4 5 7	8 9 10
40	6021	6031	6042	6053	6064	6075	6085	6096	6107	6117	1 2 3	4 5 6	8 9 10
41	6128	6138	6149	6160	6170	6180	6191	6201	6212	6222	1 2 3	4 5 6	7 8 9
42	6232	6243	6253	6263	6274	6284	6294	6304	6314	6325	1 2 3	4 5 6	7 8 9
43	6335	6345	6355	6365	6375	6385	6395	6405	6415	6425	1 2 3	4 5 6	7 8 9
44	6435	6444	6454	6464	6474	6484	6493	6503	6513	6522	1 2 3	4 5 6	7 8 9
45	6532	6542	6551	6561	6571	6580	6590	6599	6609	6618	1 2 3	4 5 6	7 8 9
46	6628	6637	6646	6656	6665	6675	6684	6693	6702	6712	1 2 3	4 5 6	7 7 8
47	6721	6730	6739	6749	6758	6767	6776	6785	6794	6803	1 2 3	4 5 5	6 7 8
48	6812	6821	6830	6839	6848	6857	6866	6875	6884	6893	1 2 3	4 4 5	6 7 8
49	6902	6911	6920	6928	6937	6946	6955	6964	6972	6981	1 2 3	4 4 5	6 7 8
50	6990	6998	7007	7016	7024	7033	7042	7050	7059	7067	1 2 3	3 4 5	6 7 8
51	7076	7084	7093	7101	7110	7118	7126	7135	7143	7152	1 2 3	3 4 5	6 7 8
52	7160	7168	7177	7185	7193	7202	7210	7218	7226	7235	1 2 2	3 4 5	6 7 7
53	7243	7251	7259	7267	7275	7284	7292	7300	7308	7316	1 2 2	3 4 5	6 6 7
54	7324	7332	7340	7348	7356	7364	7372	7380	7388	7396	1 2 2	3 4 5	6 6 7

LOGARITHMS

	0	1	2	3	4	5	6	7	8	9	1	2	3	4	5	6	7	8	9
														Differences					
55	7404	7412	7419	7427	7435	7443	7451	7459	7466	7474	1	2	2	3	4	5	5	6	7
56	7482	7490	7497	7505	7513	7520	7528	7536	7543	7551	1	2	2	3	4	5	5	6	7
57	7559	7566	7574	7582	7589	7597	7604	7612	7619	7627	1	2	2	3	4	5	5	6	7
58	7634	7642	7649	7657	7664	7672	7679	7686	7694	7701	1	1	2	3	4	4	5	6	7
59	7709	7716	7723	7731	7738	7745	7752	7760	7767	7774	1	1	2	3	4	4	5	6	7
60	7782	7789	7796	7803	7810	7818	7825	7832	7839	7846	1	1	2	3	4	4	5	6	6
61	7853	7860	7868	7875	7882	7889	7896	7903	7910	7917	1	1	2	3	4	4	5	6	6
62	7924	7931	7938	7945	7952	7959	7966	7973	7980	7987	1	1	2	3	3	4	5	6	6
63	7993	8000	8007	8014	8021	8028	8035	8041	8048	8055	1	1	2	3	3	4	5	5	6
64	8062	8069	8075	8082	8089	8096	8102	8109	8116	8122	1	1	2	3	3	4	5	5	6
65	8129	8136	8142	8149	8156	8162	8169	8176	8182	8189	1	1	2	3	3	4	5	5	6
66	8195	8202	8209	8215	8222	8228	8235	8241	8248	8254	1	1	2	3	3	4	5	5	6
67	8261	8267	8274	8280	8287	8293	8299	8306	8312	8319	1	1	2	3	3	4	5	5	6
68	8325	8331	8338	8344	8351	8357	8363	8370	8376	8382	1	1	2	3	3	4	4	5	6
69	8388	8395	8401	8407	8414	8420	8426	8432	8439	8445	1	1	2	2	3	4	4	5	6
70	8451	8457	8463	8470	8476	8482	8488	8494	8500	8506	1	1	2	2	3	4	4	5	6
71	8513	8519	8525	8531	8537	8543	8549	8555	8561	8567	1	1	2	2	3	4	4	5	5
72	8573	8579	8585	8591	8597	8603	8609	8615	8621	8627	1	1	2	2	3	4	4	5	5
73	8633	8639	8645	8651	8657	8663	8669	8675	8681	8686	1	1	2	2	3	4	4	5	5
74	8692	8698	8704	8710	.8716	8722	8727	8733	8739	8745	1	1	2	2	3	4	4	5	5
75	8751	8756	8762	8768	8774	8779	8785	8791	8797	8802	1	1	2	2	3	3	4	5	5
76	8808	8814	8820	8825	8831	8837	8842	8848	8854	8859	1	1	2	2	3	3	4	5	5
77	8865	8871	8876	8882	8887	8893	8899	8904	8910	8915	1	1	2	2	3	3	4	4	5
78	8921	8927	8932	8938	8943	8949	8954	8960	8965	8971	1	1	2	2	3	3	4	4	5
79	8976	8982	8987	8993	8998	9004	9009	9015	9020	9025	1	1	2	2	3	3	4	4	5
80	9031	9036	9042	9047	9053	9058	9063	9069	9074	9079	1	1	2	2	3	3	4	4	5
81	9085	9090	9096	9101	9106	9112	9117	9122	9128	9133	1	1	2	2	3	3	4	4	5
82	9138	9143	9149	9154	9159	9165	9170	9175	9180	9186	1	1	2	2	3	3	4	4	5
83	9191	9196	9201	9206	9212	9217	9222	9227	9232	9238	1	1	2	2	3	3	4	4	5
84	9243	9248	9253	9258	9263	9269	9274	9279	9284	9289	1	1	2	2	3	3	4	4	5
85	9294	9299	9304	9309	9315	9320	9325	9330	9335	9340	1	1	2	2	3	3	4	4	5
86	9345	9350	9355	9360	9365	9370	9375	9380	9385	9390	1	1	2	2	3	3	4	4	5
87	9395	9400	9405	9410	9415	9420	9425	9430	9435	9440	0	1	1	2	2	3	3	4	4
88	9445	9450	9455	9460	9465	9469	9474	9479	9484	9489	0	1	1	2	2	3	3	4	4
89	9494	9499	9504	9509	9513	9518	9523	9528	9533	9538	0	1	1	2	2	3	3	4	4
90	9542	9547	9552	9557	9562	9566	9571	9576	9581	9586	0	1	1	2	2	3	3	4	4
91	9590	9595	9600	9605	9609	9614	9619	9624	9628	9633	0	1	1	2	2	3	3	4	4
92	9638	9643	9647	9652	9657	9661	9666	9671	9675	9680	0	1	1	2	2	3	3	4	4
93	9685	9689	9694	9699	9703	9708	9713	9717	9722	9727	0	1	1	2	2	3	3	4	4
94	9731	9736	9741	9745	9750	9754	9759	9763	9768	9773	0	1	1	2	2	3	3	4	4
95	9777	9782	9786	9791	9795	9800	9805	9809	9814	9818	0	1	1	2	2	3	3	4	4
96	9823	9827	9832	9836	9841	9845	9850	9854	9859	9863	0	1	1	2	2	3	3	4	4
97	9868	9872	9877	9881	9886	9890	9894	9899	9903	9908	0	1	1	2	2	3	3	4	4
98	9912	9917	9921	9926	9930	9934	9939	9943	9948	9952	0	1	1	2	2	3	3	4	4
99	9956	9961	9965	9969	9974	9978	9983	9987	9991	9996	0	1	1	2	2	3	3	3	4

x

	0	1	2	3	4	5	6	7	8	9	Differences								
											1	2	3	4	5	6	7	8	9
·00	1000	1002	1005	1007	1009	1012	1014	1016	1019	1021	0	0	1	1	1	1	2	2	2
·01	1023	1026	1028	1030	1033	1035	1038	1040	1042	1045	0	0	1	1	1	1	2	2	2
·02	1047	1050	1052	1054	1057	1059	1062	1064	1067	1069	0	0	1	1	1	1	2	2	2
·03	1072	1074	1076	1079	1081	1084	1086	1089	1091	1094	0	0	1	1	1	1	2	2	2
·04	1096	1099	1102	1104	1107	1109	1112	1114	1117	1119	0	1	1	1	1	2	2	2	2
·05	1122	1125	1127	1130	1132	1135	1138	1140	1143	1146	0	1	1	1	1	2	2	2	2
·06	1148	1151	1153	1156	1159	1161	1164	1167	1169	1172	0	1	1	1	1	2	2	2	2
·07	1175	1178	1180	1183	1186	1189	1191	1194	1197	1199	0	1	1	1	1	2	2	2	2
·08	1202	1205	1208	1211	1213	1216	1219	1222	1225	1227	0	1	1	1	1	2	2	2	2
·09	1230	1233	1236	1239	1242	1245	1247	1250	1253	1256	0	1	1	1	1	2	2	2	3
·10	1259	1262	1265	1268	1271	1274	1276	1279	1282	1285	0	1	1	1	1	2	2	2	3
·11	1288	1291	1294	1297	1300	1303	1306	1309	1312	1315	0	1	1	1	2	2	2	2	3
·12	1318	1321	1324	1327	1330	1334	1337	1340	1343	1346	0	1	1	1	2	2	2	3	3
·13	1349	1352	1355	1358	1361	1365	1368	1371	1374	1377	0	1	1	1	2	2	2	3	3
·14	1380	1384	1387	1390	1393	1396	1400	1403	1406	1409	0	1	1	1	2	2	2	3	3
·15	1413	1416	1419	1422	1426	1429	1432	1435	1439	1442	0	1	1	1	2	2	2	3	3
·16	1445	1449	1452	1455	1459	1462	1466	1469	1472	1476	0	1	1	1	2	2	2	3	3
·17	1479	1483	1486	1489	1493	1496	1500	1503	1507	1510	0	1	1	1	2	2	2	3	3
·18	1514	1517	1521	1524	1528	1531	1535	1538	1542	1545	0	1	1	1	2	2	2	3	3
·19	1549	1552	1556	1560	1563	1567	1570	1574	1578	1581	0	1	1	1	2	2	2	3	3
·20	1585	1589	1592	1596	1600	1603	1607	1611	1614	1618	0	1	1	1	2	2	3	3	3
·21	1622	1626	1629	1633	1637	1641	1644	1648	1652	1656	0	1	1	2	2	2	3	3	3
·22	1660	1663	1667	1671	1675	1679	1683	1687	1690	1694	0	1	1	2	2	2	3	3	3
·23	1698	1702	1706	1710	1714	1718	1722	1726	1730	1734	0	1	1	2	2	2	3	3	4
·24	1738	1742	1746	1750	1754	1758	1762	1766	1770	1774	0	1	1	2	2	2	3	3	4
·25	1778	1782	1786	1791	1795	1799	1803	1807	1811	1816	0	1	1	2	2	2	3	3	4
·26	1820	1824	1828	1832	1837	1841	1845	1849	1854	1858	0	1	1	2	2	3	3	3	4
·27	1862	1866	1871	1875	1879	1884	1888	1892	1897	1901	0	1	1	2	2	3	3	4	4
·28	1905	1910	1914	1919	1923	1928	1932	1936	1941	1945	0	1	1	2	2	3	3	4	4
·29	1950	1954	1959	1963	1968	1972	1977	1982	1986	1991	0	1	1	2	2	3	3	4	4
·30	1995	2000	2004	2009	2014	2018	2023	2028	2032	2037	0	1	1	2	2	3	3	4	4
·31	2042	2046	2051	2056	2061	2065	2070	2075	2080	2084	0	1	1	2	2	3	3	4	4
·32	2089	2094	2099	2104	2109	2113	2118	2123	2128	2133	0	1	1	2	2	3	3	4	4
·33	2138	2143	2148	2153	2158	2163	2168	2173	2178	2183	0	1	1	2	2	3	3	4	4
·34	2188	2193	2198	2203	2208	2213	2218	2223	2228	2234	1	1	2	2	3	3	4	4	5
·35	2239	2244	2249	2254	2259	2265	2270	2275	2280	2286	1	1	2	2	3	3	4	4	5
·36	2291	2296	2301	2307	2312	2317	2323	2328	2333	2339	1	1	2	2	3	3	4	4	5
·37	2344	2350	2355	2360	2366	2371	2377	2382	2388	2393	1	1	2	2	3	3	4	4	5
·38	2399	2404	2410	2415	2421	2427	2432	2438	2443	2449	1	1	2	2	3	3	4	4	5
·39	2455	2460	2466	2472	2477	2483	2489	2495	2500	2506	1	1	2	2	3	3	4	5	5
·40	2512	2518	2523	2529	2535	2541	2547	2553	2559	2564	1	1	2	2	3	4	4	5	5
·41	2570	2576	2582	2588	2594	2600	2606	2612	2618	2624	1	1	2	2	3	4	4	5	5
·42	2630	2636	2642	2649	2655	2661	2667	2673	2679	2685	1	1	2	2	3	4	4	5	6
·43	2692	2698	2704	2710	2716	2723	2729	2735	2742	2748	1	1	2	3	3	4	4	5	6
·44	2754	2761	2767	2773	2780	2786	2793	2799	2805	2812	1	1	2	3	3	4	4	5	6
·45	2818	2825	2831	2838	2844	2851	2858	2864	2871	2877	1	1	2	3	3	4	5	5	6
·46	2884	2891	2897	2904	2911	2917	2924	2931	2938	2944	1	1	2	3	3	4	5	5	6
·47	2951	2958	2965	2972	2979	2985	2992	2999	3006	3013	1	1	2	3	3	4	5	5	6
·48	3020	3027	3034	3041	3048	3055	3062	3069	3076	3083	1	1	2	3	4	4	5	6	6
·49	3090	3097	3105	3112	3119	3126	3133	3141	3148	3155	1	1	2	3	4	4	5	6	6

ANTILOGARITHMS

	0	1	2	3	4	5	6	7	8	9	1	2	3	4	5	6	7	8	9
														Differences					
·50	3162	3170	3177	3184	3192	3199	3206	3214	3221	3228	1	1	2	3	4	4	5	6	7
·51	3236	3243	3251	3258	3266	3273	3281	3289	3296	3304	1	2	2	3	4	5	5	6	7
·52	3311	3319	3327	3334	3342	3350	3357	3365	3373	3381	1	2	2	3	4	5	5	6	7
·53	3388	3396	3404	3412	3420	3428	3436	3443	3451	3459	1	2	2	3	4	5	6	6	7
·54	3467	3475	3483	3491	3499	3508	3516	3524	3532	3540	1	2	2	3	4	5	6	6	7
·55	3548	3556	3565	3573	3581	3589	3597	3606	3614	3622	1	2	2	3	4	5	6	7	7
·56	3631	3639	3648	3656	3664	3673	3681	3690	3698	3707	1	2	3	3	4	5	6	7	8
·57	3715	3724	3733	3741	3750	3758	3767	3776	3784	3793	1	2	3	3	4	5	6	7	8
·58	3802	3811	3819	3828	3837	3846	3855	3864	3873	3882	1	2	3	4	4	5	6	7	8
·59	3890	3899	3908	3917	3926	3936	3945	3954	3963	3972	1	2	3	4	5	5	6	7	8
·60	3981	3990	3999	4009	4018	4027	4036	4046	4055	4064	1	2	3	4	5	6	6	7	8
·61	4074	4083	4093	4102	4111	4121	4130	4140	4150	4159	1	2	3	4	5	6	7	8	9
·62	4169	4178	4188	4198	4207	4217	4227	4236	4246	4256	1	2	3	4	5	6	7	8	9
·63	4266	4276	4285	4295	4305	4315	4325	4335	4345	4355	1	2	3	4	5	6	7	8	9
·64	4365	4375	4385	4395	4406	4416	4426	4436	4446	4457	1	2	3	4	5	6	7	8	9
·65	4467	4477	4487	4498	4508	4519	4529	4539	4550	4560	1	2	3	4	5	6	7	8	9
·66	4571	4581	4592	4603	4613	4624	4634	4645	4656	4667	1	2	3	4	5	6	7	9	10
·67	4677	4688	4699	4710	4721	4732	4742	4753	4764	4775	1	2	3	4	5	7	8	9	10
·68	4786	4797	4808	4819	4831	4842	4853	4864	4875	4887	1	2	3	4	6	7	8	9	10
·69	4898	4909	4920	4932	4943	4955	4966	4977	4989	5000	1	2	3	5	6	7	8	9	10
·70	5012	5023	5035	5047	5058	5070	5082	5093	5105	5117	1	2	4	5	6	7	8	9	11
·71	5129	5140	5152	5164	5176	5188	5200	5212	5224	5236	1	2	4	5	6	7	8	10	11
·72	5248	5260	5272	5284	5297	5309	5321	5333	5346	5358	1	2	4	5	6	7	9	10	11
·73	5370	5383	5395	5408	5420	5433	5445	5458	5470	5483	1	3	4	5	6	8	9	10	11
·74	5495	5508	5521	5534	5546	5559	5572	5585	5598	5610	1	3	4	5	6	8	9	10	12
·75	5623	5636	5649	5662	5675	5689	5702	5715	5728	5741	1	3	4	5	7	8	9	10	12
·76	5754	5768	5781	5794	5808	5821	5834	5848	5861	5875	1	3	4	5	7	8	9	11	12
·77	5888	5902	5916	5929	5943	5957	5970	5984	5998	6012	1	3	4	5	7	8	10	11	12
·78	6026	6039	6053	6067	6081	6095	6109	6124	6138	6152	1	3	4	6	7	8	10	11	13
·79	6166	6180	6194	6209	6223	6237	6252	6266	6281	6295	1	3	4	6	7	9	10	11	13
·80	6310	6324	6339	6353	6368	6383	6397	6412	6427	6442	1	3	4	6	7	9	10	12	13
·81	6457	6471	6486	6501	6516	6531	6546	6561	6577	6592	2	3	5	6	8	9	11	12	14
·82	6607	6622	6637	6653	6668	6683	6699	6714	6730	6745	2	3	5	6	8	9	11	12	14
·83	6761	6776	6792	6808	6823	6839	6855	6871	6887	6902	2	3	5	6	8	9	11	13	14
·84	6918	6934	6950	6966	6982	6998	7015	7031	7047	7063	2	3	5	6	8	10	11	13	15
·85	7079	7096	7112	7129	7145	7161	7178	7194	7211	7228	2	3	5	7	8	10	12	13	15
·86	7244	7261	7278	7295	7311	7328	7345	7362	7379	7396	2	3	5	7	8	10	12	13	15
·87	7413	7430	7447	7464	7482	7499	7516	7534	7551	7568	2	3	5	7	9	10	12	14	16
·88	7586	7603	7621	7638	7656	7674	7691	7709	7727	7745	2	4	5	7	9	11	12	14	16
·89	7762	7780	7798	7816	7834	7852	7870	7889	7907	7925	2	4	5	7	9	11	13	14	16
·90	7943	7962	7980	7998	8017	8035	8054	8072	8091	8110	2	4	6	7	9	11	13	15	17
·91	8128	8147	8166	8185	8204	8222	8241	8260	8279	8299	2	4	6	8	9	11	13	15	17
·92	8318	8337	8356	8375	8395	8414	8433	8453	8472	8492	2	4	6	8	10	12	14	15	17
·93	8511	8531	8551	8570	8590	8610	8630	8650	8670	8690	2	4	6	8	10	12	14	16	18
·94	8710	8730	8750	8770	8790	8810	8831	8851	8872	8892	2	4	6	8	10	12	14	16	18
·95	8913	8933	8954	8974	8995	9016	9036	9057	9078	9099	2	4	6	8	10	12	15	17	19
·96	9120	9141	9162	9183	9204	9226	9247	9268	9290	9311	2	4	6	8	11	13	15	17	19
·97	9333	9354	9376	9397	9419	9441	9462	9484	9506	9528	2	4	7	9	11	13	15	17	20
·98	9550	9572	9594	9616	9638	9661	9683	9705	9727	9750	2	4	7	9	11	13	16	18	20
·99	9772	9795	9817	9840	9863	9886	9908	9931	9954	9977	2	5	7	9	11	14	16	18	20

	0′	6′	12′	18′	24′	30′	36′	42′	48′	54′	1′	2′	3′	4′	5′
												Differences			
0°	·0000	0017	0035	0052	0070	0087	0105	0122	0140	0157	3	6	9	12	15
1	·0175	0192	0209	0227	0244	0262	0279	0297	0314	0332	3	6	9	12	15
2	·0349	0366	0384	0401	0419	0436	0454	0471	0488	0506	3	6	9	12	15
3	·0523	0541	0558	0576	0593	0610	0628	0645	0663	0680	3	6	9	12	15
4	·0698	0715	0732	0750	0767	0785	0802	0819	0837	0854	3	6	9	12	14
5	·0872	0889	0906	0924	0941	0958	0976	0993	1011	1028	3	6	9	12	14
6	·1045	1063	1080	1097	1115	1132	1149	1167	1184	1201	3	6	9	12	14
7	·1219	1236	1253	1271	1288	1305	1323	1340	1357	1374	3	6	9	12	14
8	·1392	1409	1426	1444	1461	1478	1495	1513	1530	1547	3	6	9	12	14
9	·1564	1582	1599	1616	1633	1650	1668	1685	1702	1719	3	6	9	12	14
10	·1736	1754	1771	1788	1805	1822	1840	1857	1874	1891	3	6	9	12	14
11	·1908	1925	1942	1959	1977	1994	2011	2028	2045	2062	3	6	9	11	14
12	·2079	2096	2113	2130	2147	2164	2181	2198	2215	2233	3	6	9	11	14
13	·2250	2267	2284	2300	2317	2334	2351	2368	2385	2402	3	6	8	11	14
14	·2419	2436	2453	2470	2487	2504	2521	2538	2554	2571	3	6	8	11	14
15	·2588	2605	2622	2639	2656	2672	2689	2706	2723	2740	3	6	8	11	14
16	·2756	2773	2790	2807	2823	2840	2857	2874	2890	2907	3	6	8	11	14
17	·2924	2940	2957	2974	2990	3007	3024	3040	3057	3074	3	6	8	11	14
18	·3090	3107	3123	3140	3156	3173	3190	3206	3223	3239	3	6	8	11	14
19	·3256	3272	3289	3305	3322	3338	3355	3371	3387	3404	3	5	8	11	14
20	·3420	3437	3453	3469	3486	3502	3518	3535	3551	3567	3	5	8	11	14
21	·3584	3600	3616	3633	3649	3665	3681	3697	3714	3730	3	5	8	11	14
22	·3746	3762	3778	3795	3811	3827	3843	3859	3875	3891	3	5	8	11	13
23	·3907	3923	3939	3955	3971	3987	4003	4019	4035	4051	3	5	8	11	13
24	·4067	4083	4099	4115	4131	4147	4163	4179	4195	4210	3	5	8	11	13
25	·4226	4242	4258	4274	4289	4305	4321	4337	4352	4368	3	5	8	11	13
26	·4384	4399	4415	4431	4446	4462	4478	4493	4509	4524	3	5	8	10	13
27	·4540	4555	4571	4586	4602	4617	4633	4648	4664	4679	3	5	8	10	13
28	·4695	4710	4726	4741	4756	4772	4787	4802	4818	4833	3	5	8	10	13
29	·4848	4863	4879	4894	4909	4924	4939	4955	4970	4985	3	5	8	10	13
30	·5000	5015	5030	5045	5060	5075	5090	5105	5120	5135	3	5	8	10	13
31	·5150	5165	5180	5195	5210	5225	5240	5255	5270	5284	2	5	7	10	12
32	·5299	5314	5329	5344	5358	5373	5388	5402	5417	5432	2	5	7	10	12
33	·5446	5461	5476	5490	5505	5519	5534	5548	5563	5577	2	5	7	10	12
34	·5592	5606	5621	5635	5650	5664	5678	5693	5707	5721	2	5	7	10	12
35	·5736	5750	5764	5779	5793	5807	5821	5835	5850	5864	2	5	7	9	12
36	·5878	5892	5906	5920	5934	5948	5962	5976	5990	6004	2	5	7	9	12
37	·6018	6032	6046	6060	6074	6088	6101	6115	6129	6143	2	5	7	9	12
38	·6157	6170	6184	6198	6211	6225	6239	6252	6266	6280	2	5	7	9	11
39	·6293	6307	6320	6334	6347	6361	6374	6388	6401	6414	2	4	7	9	11
40	·6428	6441	6455	6468	6481	6494	6508	6521	6534	6547	2	4	7	9	11
41	·6561	6574	6587	6600	6613	6626	6639	6652	6665	6678	2	4	7	9	11
42	·6691	6704	6717	6730	6743	6756	6769	6782	6794	6807	2	4	6	9	11
43	·6820	6833	6845	6858	6871	6884	6896	6909	6921	6934	2	4	6 ·	8	11
44	·6947	6959	6972	6984	6997	7009	7022	7034	7046	7059	2	4	6	8	10

See note (c) at p. 329.

	0'	6'	12'	18'	24'	30'	36'	42'	48'	54'	1'	2'	3'	4'	5'
											Differences				
45°	·7071	7083	7096	7108	7120	7133	7145	7157	7169	7181	2	4	6	8	10
46	·7193	7206	7218	7230	7242	7254	7266	7278	7290	7302	2	4	6	8	10
47	·7314	7325	7337	7349	7361	7373	7385	7396	7408	7420	2	4	6	8	10
48	·7431	7443	7455	7466	7478	7490	7501	7513	7524	7536	2	4	6	8	10
49	·7547	7559	7570	7581	7593	7604	7615	7627	7638	7649	2	4	6	8	9
50	·7660	7672	7683	7694	7705	7716	7727	7738	7749	7760	2	4	6	7	9
51	·7771	7782	7793	7804	7815	7826	7837	7848	7859	7869	2	4	5	7	9
52	·7880	7891	7902	7912	7923	7934	7944	7955	7965	7976	2	4	5	7	9
53	·7986	7997	8007	8018	8028	8039	8049	8059	8070	8080	2	3	5	7	9
54	·8090	8100	8111	8121	8131	8141	8151	8161	8171	8181	2	3	5	7	8
55	·8192	8202	8211	8221	8231	8241	8251	8261	8271	8281	2	3	5	7	8
56	·8290	8300	8310	8320	8329	8339	8348	8358	8368	8377	2	3	5	6	8
57	·8387	8396	8406	8415	8425	8434	8443	8453	8462	8471	2	3	5	6	8
58	·8480	8490	8499	8508	8517	8526	8536	8545	8554	8563	2	3	5	6	8
59	·8572	8581	8590	8599	8607	8616	8625	8634	8643	8652	1	3	4	6	7
60	·8660	8669	8678	8686	8695	8704	8712	8721	8729	8738	1	3	4	6	7
61	·8746	8755	8763	8771	8780	8788	8796	8805	8813	8821	1	3	4	6	7
62	·8829	8838	8846	8854	8862	8870	8878	8886	8894	8902	1	3	4	5	7
63	·8910	8918	8926	8934	8942	8949	8957	8965	8973	8980	1	3	4	5	6
64	·8988	8996	9003	9011	9018	9026	9033	9041	9048	9056	1	3	4	5	6
65	·9063	9070	9078	9085	9092	9100	9107	9114	9121	9128	1	2	4	5	6
66	·9135	9143	9150	9157	9164	9171	9178	9184	9191	9198	1	2	3	5	6
67	·9205	9212	9219	9225	9232	9239	9245	9252	9259	9265	1	2	3	4	6
68	·9272	9278	9285	9291	9298	9304	9311	9317	9323	9330	1	2	3	4	5
69	·9336	9342	9348	9354	9361	9367	9373	9379	9385	9391	1	2	3	4	5
70	·9397	9403	9409	9415	9421	9426	9432	9438	9444	9449	1	2	3	4	5
71	·9455	9461	9466	9472	9478	9483	9489	9494	9500	9505	1	2	3	4	5
72	·9511	9516	9521	9527	9532	9537	9542	9548	9553	9558	1	2	3	3	4
73	·9563	9568	9573	9578	9583	9588	9593	9598	9603	9608	1	2	2	3	4
74	·9613	9617	9622	9627	9632	9636	9641	9646	9650	9655	1	2	2	3	4
75	·9659	9664	9668	9673	9677	9681	9686	9690	9694	9699	1	1	2	3	4
76	·9703	9707	9711	9715	9720	9724	9728	9732	9736	9740	1	1	2	3	3
77	·9744	9748	9751	9755	9759	9763	9767	9770	9774	9778	1	1	2	3	3
78	·9781	9785	9789	9792	9796	9799	9803	9806	9810	9813	1	1	2	2	3
79	·9816	9820	9823	9826	9829	9833	9836	9839	9842	9845	1	1	2	2	3
80	·9848	9851	9854	9857	9860	9863	9866	9869	9871	9874	0	1	1	2	2
81	·9877	9880	9882	9885	9888	9890	9893	9895	9898	9900	0	1	1	2	2
82	·9903	9905	9907	9910	9912	9914	9917	9919	9921	9923	0	1	1	2	2
83	·9925	9928	9930	9932	9934	9936	9938	9940	9942	9943	0	1	1	1	2
84	·9945	9947	9949	9951	9952	9954	9956	9957	9959	9960	0	1	1	1	2
85	·9962	9963	9965	9966	9968	9969	9971	9972	9973	9974	0	0	1	1	1
86	·9976	9977	9979	9980	9980	9981	9982	9983	9984	9985	0	0	1	1	1
87	·9986	9987	9988	9989	9990	9990	9991	9992	9993	9993	0	0	0	1	1
88	·9994	9995	9995	9996	9996	9997	9997	9997	9998	9998	0	0	0	0	0
89	·9998	9999	9999	9999	9999	0000	0000	0000	0000	0000	0	0	0	0	0

Subtract differences

	0′	6′	12′	18′	24′	30′	36′	42′	48′	54′	1′	2′	3′	4′	5′
											Differences				
0°	1·0000	0000	0000	0000	0000	0000	9999	9999	9999	9999	0	0	0	0	0
1	·9998	9998	9998	9997	9997	9997	9996	9996	9995	9995	0	0	0	0	0
2	·9994	9993	9993	9992	9991	9990	9990	9989	9988	9987	0	0	0	1	1
3	·9986	9985	9984	9983	9982	9981	9980	9979	9978	9977	0	0	1	1	1
4	·9976	9974	9973	9972	9971	9969	9968	9966	9965	9963	0	0	1	1	1
5	·9962	9960	9959	9957	9956	9954	9952	9951	9949	9947	0	1	1	1	2
6	·9945	9943	9942	9940	9938	9936	9934	9932	9930	9928	0	1	1	1	2
7	·9925	9923	9921	9919	9917	9914	9912	9910	9907	9905	0	1	1	2	2
8	·9903	9900	9898	9895	9893	9890	9888	9885	9882	9880	0	1	1	2	2
9	·9877	9874	9871	9869	9866	9863	9860	9857	9854	9851	0	1	1	2	2
10	·9848	9845	9842	9839	9836	9833	9829	9826	9823	9820	1	1	2	2	3
11	·9816	9813	9810	9806	9803	9799	9796	9792	9789	9785	1	1	2	2	3
12	·9781	9778	9774	9770	9767	9763	9759	9755	9751	9748	1	1	2	3	3
13	·9744	9740	9736	9732	9728	9724	9720	9715	9711	9707	1	1	2	3	3
14	·9703	9699	9694	9690	9686	9681	9677	9673	9668	9664	1	1	2	3	4
15	·9659	9655	9650	9646	9641	9636	9632	9627	9622	9617	1	2	2	3	4
16	·9613	9608	9603	9598	9593	9588	9583	9578	9573	9568	1	2	2	3	4
17	·9563	9558	9553	9548	9542	9537	9532	9527	9521	9516	1	2	3	3	4
18	·9511	9505	9500	9494	9489	9483	9478	9472	9466	9461	1	2	3	4	5
19	·9455	9449	9444	9438	9432	9426	9421	9415	9409	9403	1	2	3	4	5
20	·9397	9391	9385	9379	9373	9367	9361	9354	9348	9342	1	2	3	4	5
21	·9336	9330	9323	9317	9311	9304	9298	9291	9285	9278	1	2	3	4	5
22	·9272	9265	9259	9252	9245	9239	9232	9225	9219	9212	1	2	3	4	6
23	·9205	9198	9191	9184	9178	9171	9164	9157	9150	9143	1	2	3	5	6
24	·9135	9128	9121	9114	9107	9100	9092	9085	9078	9070	1	2	4	5	6
25	·9063	9056	9048	9041	9033	9026	9018	9011	9003	8996	1	3	4	5	6
26	·8988	8980	8973	8965	8957	8949	8942	8934	8926	8918	1	3	4	5	6
27	·8910	8902	8894	8886	8878	8870	8862	8854	8846	8838	1	3	4	5	7
28	·8829	8821	8813	8805	8796	8788	8780	8771	8763	8755	1	3	4	6	7
29	·8746	8738	8729	8721	8712	8704	8695	8686	8678	8669	1	3	4	6	7
30	·8660	8652	8643	8634	8625	8616	8607	8599	8590	8581	1	3	4	6	7
31	·8572	8563	8554	8545	8536	8526	8517	8508	8499	8490	2	3	5	6	8
32	·8480	8471	8462	8453	8443	8434	8425	8415	8406	8396	2	3	5	6	8
33	·8387	8377	8368	8358	8348	8339	8329	8320	8310	8300	2	3	5	6	8
34	·8290	8281	8271	8261	8251	8241	8231	8221	8211	8202	2	3	5	7	8
35	·8192	8181	8171	8161	8151	8141	8131	8121	8111	8100	2	3	5	7	8
36	·8090	8080	8070	8059	8049	8039	8028	8018	8007	7997	2	3	5	7	9
37	·7986	7976	7965	7955	7944	7934	7923	7912	7902	7891	2	4	5	7	9
38	·7880	7869	7859	7848	7837	7826	7815	7804	7793	7782	2	4	5	7	9
39	·7771	7760	7749	7738	7727	7716	7705	7694	7683	7672	2	4	6	7	9
40	·7660	7649	7638	7627	7615	7604	7593	7581	7570	7559	2	4	6	8	9
41	·7547	7536	7524	7513	7501	7490	7478	7466	7455	7443	2	4	6	8	10
42	·7431	7420	7408	7396	7385	7373	7361	7349	7337	7325	2	4	6	8	10
43	·7314	7302	7290	7278	7266	7254	7242	7230	7218	7206	2	4	6	8	10
44	·7193	7181	7169	7157	7145	7133	7120	7108	7096	7083	2	4	6	8	10

See note (c) at p. 329.

NATURAL COSINES

Subtract differences

	0′	6′	12′	18′	24′	30′	36′	42′	48′	54′	Differences 1′	2′	3′	4′	5′
45°	·7071	7059	7046	7034	7022	7009	6997	6984	6972	6959	2	4	6	8	10
46	·6947	6934	6921	6909	6896	6884	6871	6858	6845	6833	2	4	6	8	11
47	·6820	6807	6794	6782	6769	6756	6743	6730	6717	6704	2	4	6	9	11
48	·6691	6678	6665	6652	6639	6626	6613	6600	6587	6574	2	4	7	9	11
49	·6561	6547	6534	6521	6508	6494	6481	6468	6455	6441	2	4	7	9	11
50	·6428	6414	6401	6388	6374	6361	6347	6334	6320	6307	2	4	7	9	11
51	·6293	6280	6266	6252	6239	6225	6211	6198	6184	6170	2	5	7	9	11
52	·6157	6143	6129	6115	6101	6088	6074	6060	6046	6032	2	5	7	9	12
53	·6018	6004	5990	5976	5962	5948	5934	5920	5906	5892	2	5	7	9	12
54	·5878	5864	5850	5835	5821	5807	5793	5779	5764	5750	2	5	7	9	12
55	·5736	5721	5707	5693	5678	5664	5650	5635	5621	5606	2	5	7	10	12
56	·5592	5577	5563	5548	5534	5519	5505	5490	5476	5461	2	5	7	10	12
57	·5446	5432	5417	5402	5388	5373	5358	5344	5329	5314	2	5	7	10	12
58	·5299	5284	5270	5255	5240	5225	5210	5195	5180	5165	2	5	7	10	12
59	·5150	5135	5120	5105	5090	5075	5060	5045	5030	5015	3	5	8	10	13
60	·5000	4985	4970	4955	4939	4924	4909	4894	4879	4863	3	5	8	10	13
61	·4848	4833	4818	4802	4787	4772	4756	4741	4726	4710	3	5	8	10	13
62	·4695	4679	4664	4648	4633	4617	4602	4586	4571	4555	3	5	8	10	13
63	·4540	4524	4509	4493	4478	4462	4446	4431	4415	4399	3	5	8	10	13
64	·4384	4368	4352	4337	4321	4305	4289	4274	4258	4242	3	5	8	11	13
65	·4226	4210	4195	4179	4163	4147	4131	4115	4099	4083	3	5	8	11	13
66	·4067	4051	4035	4019	4003	3987	3971	3955	3939	3923	3	5	8	11	13
67	·3907	3891	3875	3859	3843	3827	3811	3795	3778	3762	3	5	8	11	13
68	·3746	3730	3714	3697	3681	3665	3649	3633	3616	3600	3	5	8	11	14
69	·3584	3567	3551	3535	3518	3502	3486	3469	3453	3437	3	5	8	11	14
70	·3420	3404	3387	3371	3355	3338	3322	3305	3289	3272	3	5	8	11	14
71	·3256	3239	3223	3206	3190	3173	3156	3140	3123	3107	3	6	8	11	14
72	·3090	3074	3057	3040	3024	3007	2990	2974	2957	2940	3	6	8	11	14
73	·2924	2907	2890	2874	2857	2840	2823	2807	2790	2773	3	6	8	11	14
74	·2756	2740	2723	2706	2689	2672	2656	2639	2622	2605	3	6	8	11	14
75	·2588	2571	2554	2538	2521	2504	2487	2470	2453	2436	3	6	8	11	14
76	·2419	2402	2385	2368	2351	2334	2317	2300	2284	2267	3	6	8	11	14
77	·2250	2233	2215	2198	2181	2164	2147	2130	2113	2096	3	6	9	11	14
78	·2079	2062	2045	2028	2011	1994	1977	1959	1942	1925	3	6	9	11	14
79	·1908	1891	1874	1857	1840	1822	1805	1788	1771	1754	3	6	9	11	14
80	·1736	1719	1702	1685	1668	1650	1633	1616	1599	1582	3	6	9	12	14
81	·1564	1547	1530	1513	1495	1478	1461	1444	1426	1409	3	6	9	12	14
82	·1392	1374	1357	1340	1323	1305	1288	1271	1253	1236	3	6	9	12	14
83	·1219	1201	1184	1167	1149	1132	1115	1097	1080	1063	3	6	9	12	14
84	·1045	1028	1011	0993	0976	0958	0941	0924	0906	0889	3	6	9	12	14
85	·0872	0854	0837	0819	0802	0785	0767	0750	0732	0715	3	6	9	12	14
86	·0698	0680	0663	0645	0628	0610	0593	0576	0558	0541	3	6	9	12	15
87	·0523	0506	0488	0471	0454	0436	0419	0401	0384	0366	3	6	9	12	15
88	·0349	0332	0314	0297	0279	0262	0244	0227	0209	0192	3	6	9	12	15
89	·0175	0157	0140	0122	0105	0087	0070	0052	0035	0017	3	6	9	12	15

NATURAL TANGENTS

	0′	6′	12′	18′	24′	30′	36′	42′	48′	54′	Differences				
											1′	2′	3′	4′	5′
0°	·0000	0017	0035	0052	0070	0087	0105	0122	0140	0157	3	6	9	12	15
1	·0175	0192	0209	0227	0244	0262	0279	0297	0314	0332	3	6	9	12	15
2	·0349	0367	0384	0402	0419	0437	0454	0472	0489	0507	3	6	9	12	15
3	·0524	0542	0559	0577	0594	0612	0629	0647	0664	0682	3	6	9	12	15
4	·0699	0717	0734	0752	0769	0787	0805	0822	0840	0857	3	6	9	12	15
5	·0875	0892	0910	0928	0945	0963	0981	0998	1016	1033	3	6	9	12	15
6	·1051	1009	1086	1104	1122	1139	1157	1175	1192	1210	3	6	9	12	15
7	·1228	1246	1263	1281	1299	1317	1334	1352	1370	1388	3	6	9	12	15
8	·1405	1423	1441	1459	1477	1495	1512	1530	1548	1566	3	6	9	12	15
9	·1584	1602	1620	1638	1655	1673	1691	1709	1727	1745	3	6	9	12	15
10	·1763	1781	1799	1817	1835	1853	1871	1890	1908	1926	3	6	9	12	15
11	·1944	1962	1980	1908	2016	2035	2053	2071	2089	2107	3	6	9	12	15
12	·2126	2144	2162	2180	2199	2217	2235	2254	2272	2290	3	6	9	12	15
13	·2309	2327	2345	2364	2382	2401	2419	2438	2456	2475	3	6	9	12	15
14	·2493	2512	2530	2549	2568	2586	2605	2623	2642	2661	3	6	9	12	16
15	·2679	2698	2717	2736	2754	2773	2792	2811	2830	2849	3	6	9	13	16
16	·2867	2886	2905	2924	2943	2962	2981	3000	3019	3038	3	6	9	13	16
17	·3057	3076	3096	3115	3134	3153	3172	3191	3211	3230	3	6	10	13	16
18	·3249	3269	3288	3307	3327	3346	3365	3385	3404	3424	3	6	10	13	16
19	·3443	3463	3482	3502	3522	3541	3561	3581	3600	3620	3	7	10	13	16
20	·3640	3659	3679	3699	3719	3739	3759	3779	3799	3819	3	7	10	13	17
21	·3839	3859	3879	3899	3919	3939	3959	3979	4000	4020	3	7	10	13	17
22	·4040	4061	4081	4101	4122	4142	4163	4183	4204	4224	3	7	10	14	17
23	·4245	4265	4286	4307	4327	4348	4369	4390	4411	4431	3	7	10	14	17
24	·4452	4473	4494	4515	4536	4557	4578	4599	4621	4642	4	7	11	14	18
25	·4663	4684	4706	4727	4748	4770	4791	4813	4834	4856	4	7	11	14	18
26	·4877	4899	4921	4942	4964	4986	5008	5029	5051	5073	4	7	11	15	18
27	·5095	5117	5139	5161	5184	5206	5228	5250	5272	5295	4	7	11	15	18
28	·5317	5340	5362	5384	5407	5430	5452	5475	5498	5520	4	8	11	15	19
29	·5543	5566	5589	5612	5635	5658	5681	5704	5727	5750	4	8	12	15	19
30	·5774	5797	5820	5844	5867	5890	5914	5938	5961	5985	4	8	12	16	20
31	·6009	6032	6056	6080	6104	6128	6152	6176	6200	6224	4	8	12	16	20
32	·6249	6273	6297	6322	6346	6371	6395	6420	6445	6469	4	8	12	16	20
33	·6494	6519	6544	6569	6594	6619	6644	6669	6694	6720	4	8	13	17	21
34	·6745	6771	6796	6822	6847	6873	6899	6924	6950	6976	4	9	13	17	21
35	·7002	7028	7054	7080	7107	7133	7159	7186	7212	7239	4	9	13	18	22
36	·7265	7292	7319	7346	7373	7400	7427	7454	7481	7508	5	9	14	18	23
37	·7536	7563	7590	7618	7646	7673	7701	7729	7757	7785	5	9	14	18	23
38	·7813	7841	7869	7898	7926	7954	7983	8012	8040	8069	5	9	14	19	24
39	·8098	8127	8156	8185	8214	8243	8273	8302	8332	8361	5	10	15	20	24
40	·8391	8421	8451	8481	8511	8541	8571	8601	8632	8662	5	10	15	20	25
41	·8693	8724	8754	8785	8816	8847	8878	8910	8941	8972	5	10	16	21	26
42	·9004	9036	9067	9099	9131	9163	9195	9228	9260	9293	5	11	16	21	27
43	·9325	9358	9391	9424	9457	9490	9523	9556	9590	9623	6	11	17	22	28
44	·9657	9691	9725	9759	9793	9827	9861	9896	9930	9965	6	11	17	23	29

NATURAL TANGENTS

	0′	6′	12′	18′	24′	30′	36′	42′	48′	54′	Differences				
											1′	2′	3′	4′	5′
45°	1·0000	0035	0070	0105	0141	0176	0212	0247	0283	0319	6	12	18	24	30
46	1·0355	0392	0428	0464	0501	0538	0575	0612	0649	0686	6	12	18	25	31
47	1·0724	0761	0799	0837	0875	0913	0951	0990	1028	1067	6	13	19	25	32
48	1·1106	1145	1184	1224	1263	1303	1343	1383	1423	1463	7	13	20	27	33
49	1·1504	1544	1585	1626	1667	1708	1750	1792	1833	1875	7	14	21	28	34
50	1·1918	1960	2002	2045	2088	2131	2174	2218	2261	2305	7	14	22	29	36
51	1·2349	2393	2437	2482	2527	2572	2617	2662	2708	2753	8	15	23	30	38
52	1·2799	2846	2802	2938	2985	3032	3079	3127	3175	3222	8	16	24	31	39
53	1·3270	3319	3367	3416	3465	3514	3564	3613	3663	3713	8	16	25	33	41
54	1·3764	3814	3865	3916	3968	4019	4071	4124	4176	4229	9	17	26	34	43
55	1·4281	4335	4388	4442	4496	4550	4605	4650	4715	4770	9	18	27	36	45
56	1·4826	4882	4938	4994	5051	5108	5166	5224	5282	5340	10	19	29	38	48
57	1·5399	5458	5517	5577	5637	5697	5757	5818	5880	5941	10	20	30	40	50
58	1·6003	6066	6128	6191	6255	6319	6383	6447	6512	6577	11	21	32	43	53
59	1·6643	6709	6775	6842	6909	6977	7045	7113	7182	7251	11	23	34	45	56
60	1·7321	7391	7461	7532	7603	7675	7747	7820	7893	7966	12	24	36	48	60
61	1·8040	8115	8190	8265	8341	8418	8495	8572	8650	8728	13	26	38	51	64
62	1·8807	8887	8967	9047	9128	9210	9292	9375	9458	9542	14	27	41	55	68
63	1·9626	9711	9797	9883	9970	0057	0145	0233	0323	0413	15	29	44	58	73
64	2·0503	0594	0686	0778	0872	0965	1060	1155	1251	1348	16	31	47	63	78
65	2·1445	1543	1642	1742	1842	1943	2045	2148	2251	2355	17	34	51	68	85
66	2·2460	2566	2673	2781	2889	2998	3109	3220	3332	3445	18	37	55	73	92
67	2·3559	3673	3789	3906	4023	4142	4262	4383	4504	4627	20	40	60	79	99
68	2·4751	4876	5002	5129	5257	5386	5517	5649	5782	5916	22	43	65	87	108
69	2·6051	6187	6325	6464	6605	6746	6889	7034	7179	7326	24	47	71	95	119
70	2·7475	7625	7776	7929	8083	8239	8397	8556	8716	8878	26	52	78	104	131
71	2·9042	9208	9375	9544	9714	9887	0061	0237	0415	0595	29	58	87	116	145
72	3·0777	0961	1146	1334	1524	1716	1910	2106	2305	2506	32	64	96	129	161
73	3·2709	2914	3122	3332	3544	3759	3977	4197	4420	4646	36	72	108	144	180
74	3·4874	5105	5339	5576	5816	6059	6305	6554	6806	7062	41	81	122	163	204
75	3·7321	7583	7848	8118	8391	8667	8947	9232	9520	9812	46	93	139	186	232
76	4·0108	0408	0713	1022	1335	1653	1976	2303	2635	2972					
77	4·3315	3662	4015	4374	4737	5107	5483	5864	6252	6646					
78	4·7046	7453	7867	8288	8716	9152	9504	0045	0504	0970					
79	5·1446	1929	2422	2024	3435	3955	4486	5026	5578	6140					
80	5·6713	7297	7894	8502	9124	9758	0405	1066	1742	2432					
81	6·3138	3859	4596	5350	6122	6912	7720	8548	9395	0264					
82	7·1154	2066	3002	3962	4947	5958	6996	8062	9158	0285					
83	8·1443	2636	3863	5126	6427	7709	9152	0579	2052	3572					
84	9·514	9·677	9·845	10·02	10·20	10·39	10·58	10·78	10·99	11·20					
85	11·43	11·66	11·91	12·16	12·43	12·71	13·00	13·30	13·62	13·95					
86	14·30	14·67	15·06	15·46	15·89	16·35	16·83	17·34	17·89	18·46					
87	19·08	19·74	20·45	21·20	22·02	22·90	23·86	24·90	26·03	27·27					
88	28·64	30·14	31·82	33·69	35·80	38·19	40·92	44·07	47·74	52·08					
89	57·29	63·66	71·62	81·85	95·49	114·6	143·2	191·0	286·5	573·0					

Notes. (*a*) The whole number at the left of any row (up to the 84° row) is to be read with each decimal fraction in the same row.

(*b*) Where the figures in any row are printed in slightly heavier type the whole number at the left of that row is increased by 1 for those figures.

(*c*) Decimal points are to be read before each four-figure group where they are not shown.

ANSWERS

EXERCISE 1

P. 39.   1. 1, 3, 5, . . . 45, 47, 49; 2, 3, 5, 7, 11, 13, 17, 19, 23,
29, 31, 37, 41, 43, 47.

2. The sum of the first n odd numbers is n^2.

3. a^7; 3^8; 10^8; 2^6; 5^8; x^3; 7; a^6; 1; 10^6.

4. 13; 19; 24; 18; 27.   5. 14, 15; 12, 13; 15, 16; 21, 22.

6. 2648, 3332, 123,456, 1,223,748; 435, 690, 2565;
2565, 1,223,748; 690; 690, 123456, 1,223,748; 2648,
123,456; 123,456, 1,223,748.   7. 120.   8. 120.

9. 15.   10. 24.   11. 240.   12. 86,400.

EXERCISE 2

P. 49.   1. (a) $+ £10$, $- £5$; £5 in hand.   (b) $+ £4$, $+ £6$;
£10 in hand.   (c) $- 5s.$, $- 4s.$, spent 9s.   (d) $+ 14°$,
$- 8°$; temp. 6° up.   (e) $- 250$ ft., $+ 100$ ft.; 150 ft.
below first level.   (f) $- 8$ ft., $+ 7$ ft.; 1 ft. below
first level.

2. June 9; June 23; July 6; June 5; May 22.

3. (a) 25 ft. below sea-level.   (b) 8 ft. behind mark.
(c) 4° below zero.   (d) 20 m.p.h. in reverse or in the
reverse direction.

4. $+ 6$; $- 22$; $- 22$; $+ 5$; $+ 50$; $- 34$; 0 ; $+ 9$.

5. $- 52$; $+ 12$; $- 72$; $- 60$; $- 4$; $+ 4$; $- 6$.

6. 16, 49, 81, 144, 225; $- 27$, $- 64$, $- 216$.

EXERCISE 3

P. 56.   1. $\dfrac{2}{3}, \dfrac{3}{4}, \dfrac{2}{3}, \dfrac{3}{4}, \dfrac{3}{4}, \dfrac{1}{5}, \dfrac{1}{4}, \dfrac{5}{8}, \dfrac{5}{8}, \dfrac{2}{3}$.

2. $\dfrac{4}{10}, \dfrac{20}{50}; \dfrac{9}{12}, \dfrac{45}{60}; \dfrac{24}{36}, \dfrac{36}{54}$

3. 12, 20, 40, 70, 15.

EXERCISE 4

P. 59.   1. $\frac{5}{6}$; $\frac{11}{12}$; $1\frac{1}{2}$; $1\frac{1}{12}$; 2.

2. $\frac{1}{6}$; $\frac{5}{12}$; $1\frac{1}{6}$; $\frac{3}{4}$; $5\frac{5}{6}$.

3. $16\frac{5}{6}$; $15\frac{3}{4}$; $7\frac{3}{8}$; $6\frac{1}{3}$.

331

EXERCISE 5

P. 62.   1. $\dfrac{7}{12}$; $\dfrac{2}{21}$; $\dfrac{2}{3}$; $\dfrac{2}{5}$; $\dfrac{2}{9}$.

2. $7\frac{7}{8}$; $12\frac{1}{10}$; $1\frac{1}{8}$; 20; 16.

3. $1\frac{3}{4}$; $16\frac{2}{3}$; $9\frac{3}{8}$.   4. $\frac{8}{15}$; $\frac{8}{9}$; $1\frac{1}{2}$; 12; $\frac{1}{100}$.

5. $4\frac{2}{7}$; 18; $\frac{1}{6}$; 4; $2\frac{2}{3}$.

EXERCISE 6

	(a)	(b)	(c)	(d)
P. 69. 1.	14·08;	16·84;	31·7;	67.
2.	21·202;	23·494;	68·85;	50.
3.	3·92;	8·25;	·0065;	·05.
4.	12·35;	13·75;	16·5;	5·75.
6.	8·98.			

EXERCISE 7

P. 70.   1. 65; 70; 216; 207·2; 514·8; 29·38; 18·0324.

2. 3·7; 1·9; 5·64; ·95; ·323; ·294; ·00625.

EXERCISE 8

P. 81.   1. 42; 120; £10; £1.   2. 31.   3. 12.   4. 79%.

5. £612; £3 1s. 2d.   6. $6\frac{2}{3}$%.   7. 8s. 3d.

8. £5 12s. 6d.   9. 1200.   10. Too much by £1.

EXERCISE 9

P. 89.   1. 12,010; 14,034; 7905; 30,132.

2. Cross: 146, 209, 217, 278, 186. Down: 243, 180, 236, 187, 190. Check: 1036.

3. 29; 18; 117; 209; 377; 1878; 3457.

4. £29 3s. 8d.; £59 1s.; £173 5s. $9\frac{1}{2}$d.; £4394 16s. 7d.

5. £16 18s. 5d.; £1 4s. 11d.; £1 18s. $1\frac{1}{2}$d.; £40 17s. 10d.

EXERCISE 10

P. 95.   1. 768; 2166; 1296; 999; 1273.   2. 7488; 30,805; 327,900; 17,640; 69,768.

3. 15·96; 19·6; 1·5432; ·00025.   4. 24; 25; 27; 123, rem. 172; 6944, rem. 64.   5. 13·75; 3·8; 8·6; ·0725; ·003072.

EXERCISE 11

P. 96. 2nd; 3rd; 4th (try 11); 5th (try 7); seventh; last.

EXERCISE 12

P. 98. 1. 875; 4340; 3·7; 3·5; 373·5. 2. 1050; 1625; 107·5;
10,875; 700. 3. 2583; 3078; 4554; 8613; 2716;
475,524. 4. 7s. 6d.; £1 9s. 6d.; £2 6s. 3d.; £4 5s.;
£15 10s.; £18 15s.; 2½ cwt.; 3 ton 5 cwt.; £7 15s.;
£2 2s. 6d.

EXERCISE 13

P. 100. 1. ·387; 4·39; ·0245; ·0375; ·086. 2. 16·8; 138·24;
57·032; ·36. 3. 1s. 2¾d.; 8¼d.; 2s. 9d.; 14s. 8d.;
10d.; 10½d.; 15 lb. 8 oz. 4. 20·3125; 13·4375;
13·1171875.

EXERCISE 14

P. 104. 1. 60. 2. No answer. 3. 1s. 5½d. 4. 16 days. 5.
4s. 3d. 6. No answer. 7. 24 days. 8. 126 miles.
9. 8 days. 10. 200.

EXERCISE 15

P. 108. 3. (a) 40 : 7; (b) 22 : 5: (c) No answer; (d) 1 : 10,560;
(e) No answer. 4. 28 : 25. 5. 19 : 18. 6. (a) 4 ft.
8 in.; (b) 4 ft. 2 in.; (c) No. 7. 4 : 3. 8. No.

EXERCISE 16

P. 110. 1. £3 4s.; £4; £4 16s. 2. £2 12s. 6d.; £1 2s. 6d.
3. £3; £1 10s.; 15s. 4. Yes, 116 of each. 5. 15s.;
13s. 6d.; 12s.

EXERCISE 17

P. 113. 1. (a) £28; (b) £6 15s.; (c) £2 5s.
2. (a) 5s. 2½d.; (b) 5s. 3d.; (c) 5s. 3. 12s.

EXERCISE 18

P. 115. 1. 3⅛%.   2. 5⅚ (5·83)%; 6$\frac{2}{93}$ (6·02)%.
3. Gain of £1 14s. 5d.
4. 3$\frac{17}{81}$ or £3 16s. 2d. per cent.

EXERCISE 19

P. 127. 1. (a) £2 6s., £3 14s. 9d., £5 3s. 6d., 17s. 3d.;
(b) £1 14s. 9d., £2 16s. 6d., £3 18s. 3d., 13s. 1d.
2. (a) £2 9s. 2d., £2 1s., £3 1s. 6d., £1 2s. 7d.;
(b) £2 8s. 9d., £3 1s., £2 2s. 8d., 15s. 3d.
3. (a) 20·8, 15·4, 10·2, 25·3, 15·0; (b) 6·3, 3·1, 4·7, 7·3, 5·8.
4. (a) 37, 11, 33 miles; (b) 16, 7·5, 77 kilometres.
5. (a) 41 m.p.h.; (b) 30 miles; (c) 47 min.

EXERCISE 20

P. 130. 1. (a) 60 ft., 212½ ft., 90 ft., 27 ft.
(b) 23 m.p.h., 42 m.p.h., 56 m.p.h., 36½ m.p.h.
2. 10 ft., 30 ft., 55 ft., 90 ft., 135 ft.

EXERCISE 21

P. 161. 1. (a) 62 sq. ft.; (b) 28 sq. m.; (c) 176 sq. in.; (d) $4\frac{3}{7}$ sq. ft.; (e) 128 sq. cm.; (f) $19\frac{2}{7}$ sq. ft. 2. $13\frac{3}{14}$ tons.
3. 260 sq. ft. 4. 220 sq. ft.; 10,560 cu. ft. 5. (a) 366 sq. in.; (b) 183 cu. in. 6. 354 sq. in.

EXERCISE 22

P. 185. 1. 15 yd. 2. 7 miles. 3. 2 ft. 2 in. 4. 10 ft.

EXERCISE 23

P. 196. $\dfrac{5}{12}$; $\dfrac{y}{z}$; $\dfrac{3}{5}$; $\dfrac{7}{25}$; $\dfrac{c}{a}$; $\dfrac{5}{13}$; $\dfrac{q}{r}$; $\dfrac{y}{x}$; $\dfrac{c}{b}$; $\dfrac{12}{13}$; $\dfrac{24}{7}$; $\dfrac{x}{z}$; $\dfrac{p}{r}$; $\dfrac{24}{25}$; $\dfrac{4}{3}$; $\dfrac{a}{b}$; $\dfrac{4}{5}$; $\dfrac{p}{q}$.

EXERCISE 24

P. 198. 1. ·1736; ·7660; ·9848; ·2672; ·4879; ·9993; ·3976; ·6793; ·9387.
2. ·9962; ·7071; ·2924; ·9041; ·5920; ·1668; ·9529; ·7339; ·2354.
3. ·2493; 1·0000; 1·7321; ·3659; 1·1750; 5·0504; ·2385; 1·0476; 3·9371.
4. 13°; 55°; 22° 24′; 68° 42′; 35° 39′; 71° 38′.
5. 22°; 60°; 12° 42′; 70° 18′; 36° 52′; 74° 38′.
6. 40°; 45°; 30° 36′; 58° 48′; 68° 7′; 75° 58′.

ANSWERS

EXERCISE 25

P. 202. 1. $c = 5$; $A = 53° 8'$; $B = 36° 52'$.

2. $a = 12$; $A = 67° 23'$; $C = 22° 37'$.

3. $a = 7$; $A = 16° 15'$; $B = 73° 45'$.

4. $A = 60°$; $a = 13·9$; $c = 8$.

5. $A = 25°$; $b = 85·8$; $C = 94·6$.

6. $a = 335$; $B = 26° 34'$; $C = 63° 26'$.

7. $B = 36° 44'$; $b = 15·0$; $c = 20·0$.

8. $0° 46'$. 9. Yes, by 37·8 ft. 10. 15·6 ft.

11. 30° 58' W. of N.; 30° 58' W. of S. 12. 5° 43'; 5° 13'

EXERCISE 26

P. 210. 1. 864. 2. 14,080. 3. 50. 4. 2100. 5. 2; $6\frac{1}{4}$; $10\frac{1}{2}$.

6. 18,400. 7. 24. 8. 2. 9. 24,200. 10. $l = \dfrac{A}{b}$.

11. $V = \dfrac{RT}{P}$ 12. $P = \dfrac{33,000H}{LAN}$ 13. $r = \sqrt{\dfrac{A}{\pi}}$.

14. $t = \dfrac{2s}{u + v}$. 15. $R = \dfrac{V}{\pi ab} + r$.

16. (i) $u = \dfrac{1}{t}(s - \frac{1}{2} \text{ ft}^2)$; (ii) $f = \dfrac{2}{t^2}(s - ut)$.

EXERCISE 27

P. 217. 1. $5 + x + y$; $17y + a - 4$; $30 + 16l - 3t$; $3a + 4b - 6 + 5l$; $2r + 3s + 8 - 5a - 7c$; $4 - a - b$; $7 - 2x - 3y$; $y - x - 2$; $2u - 3v - 15 + 2w$; $3a - 4b - c + 11e - f$.

2. $4x + 12$; $7y - 35$; $6a + 27$; $80 - 30t$; $a^2 + 17a$; $2a^2 + 3a$; $5b^2 - ba$; $6x^2 - 45x$; $77 + 21A + 10B - 2BC$; $14A^2 + 14AB - 54 + 3B$.

3. $a^2 + 2a + 1$; $b^2 - 10b + 25$; $x^2 + 4x + 4$; $4t^2 - 12t + 9$; $25x^2 + 40x + 16$; $x^2 + 6xy + 9y^2$; $9a^2 - 30ab + 25b^2$; $x^2 - 4$; $4t^2 - 9$; $9 - 49x^2$.

5. $ab + 2a + 3b + 6$; $xy - 7x + 4y - 28$; $ab - 5a - 2b + 10$; $6 + 3a + 2b + ba$; $60 + 5t - 12x - xt$.

6. 2401; 5329; 7225; 9801; 10,609; 15,376.

EXERCISE 28

P. 226. 1. 10. 2. $-2\frac{1}{2}$. 3. -17. 4. 13. 5. $1\frac{1}{2}$. 6. -3.
7. -1. 8. $13\frac{1}{2}$. 9. 16. 10. $-4\frac{1}{2}$. 11. $-2\frac{2}{3}$.
12. 7. 13. 2. 14. 4. 15. 20. 16. 10. 17. 99.
18. 440. 19. $5\frac{1}{3}$. 20. 64. 21. 1,200,000. 22. 312.
23. 1714.

EXERCISE 29

P. 229. 1. 7. 2. 19. 3. 12. 4. 10. 5. 12. 6. 48 years, 12
years. 7. 32 miles. 8. 100. 9. 392. 10. 10 miles.
11. 7 days. 12. $4s.$; $2s.$; $9s.$

EXERCISE 30

P. 235. 1. 9, 7. 2. 3, 2. 3. 115, 5. 4. 11, 7. 5. 2, -2.
6. 20, 30. 7. 14, 11. 8. $-6, 7$. 9. $8\frac{1}{3}, 6\frac{1}{3}$. 10. $4\frac{1}{2}$,
$6\frac{3}{4}$. 11. 10, 6. 12. 7, 2. 13. 6, 3. 14. 3, 2. 15. 42,
24. 16. 10, 24.

EXERCISE 31

P. 237. 1. 51, 36. 2. 50, 40. 3. 18, 12. 4. 48, 32. 5. 30, 20.
6. $\frac{1}{4}d.$ 7. 18, 12. 8. 20. 9. 34, 16. 10. 91 men,
39 women.
11. (i) $\frac{1}{20}$, 60; (ii) 63 lb.; (iii) 120 lb.; (iv) No.
12. (i) 15, $\frac{1}{4}$; (ii) 15 in.; (iii) $7\frac{1}{2}$ in.; (iv) 4 lb.

EXERCISE 32

P. 241. 1. 1 or 2. 2. -2 or -5. 3. 2 or 4. 4. 2 or -4.
5. 2 or $\frac{1}{2}$. 6. $1\frac{1}{3}$ or $-1\frac{1}{2}$. 7. $\frac{2}{3}$ or $-2\frac{1}{2}$. 8. 7 or $\frac{1}{2}$.
9. 16 or $-1\frac{1}{2}$. 10. 7 or -8.

EXERCISE 33

P. 243. 1. 6 (or -7). 2. 9 and 17 (or -17 and -9). 3. 10
and 12. 4. (i) 210; (ii) 4230; (iii) 14. 5. 12. 6. 6.
7. 3 ft.

EXERCISE 34

P. 251. 3. (i) 76 lb.; (ii) $16\frac{1}{2}$ lb. 6. The second pair.
7. (i) 32 lb.; (ii) 196 lb.; (iii) 16 lb. 8. (i) 20·3; (ii) 5·5.
9. (i) 46·7; (ii) 4·6.
10. (i) 100 ft.; (ii) 16 ft. per sec.; (iii) 48 ft. per sec.
11. (a) £130; (b) 3 yr. 2 months.
12. The fourth.

ANSWERS

EXERCISE 35

P. 280. ·6812, ·8633, ·3979, ·8195, ·3692, ·7316, ·6794, ·8058, ·7281, ·6990, ·7782, ·8451, 1·3010, 1·6990, 1·6590, 1·9201, 1·5623, 2·5386, 2·9419, 2·3717, 1·6682, 1·9655, 3·5546, 3·4746, 5·0916, 2·9506, $\overline{1}$·7924, $\overline{1}$·7634, $\overline{1}$·7497, $\overline{1}$·5904, $\overline{1}$·9542, $\overline{2}$·5315, $\overline{2}$·8344, $\overline{3}$·5611, $\overline{5}$·9186, $\overline{2}$·7825, $\overline{3}$·1973, $\overline{9}$·4679.

EXERCISE 36

P. 281. 2·291, 3·802, 3·69, 1·945, 2·263, 5·751, 22·23, 886·1, 2154, 501600, ·2985, ·03637, ·05441, ·00367 ·0007286.

EXERCISE 37

P. 283. 1. 27·0. 2. 98·8. 3. 1550. 4. 13·2. 5. 71·6. 6. 8810. 7. 10·5. 8. 18·6. 9. 66·7. 10. 62·8. 11. 169. 12. 18·9. 13. 94,800. 14. 706. 15. 5·68. 16. 12·9. 17. 7·59. 18. 2·98.

EXERCISE 38

P. 285. 1. 13·5. 2. ·382. 3. 3·85. 4. ·000448. 5. 28·0. 6. ·730. 7. ·0679. 8. ·00202. 9. ·0176. 10. ·127. 11. ·000824. 12. ·0441. 13. ·920. 14. ·316. 15. ·255. 16. ·715. 17. ·928. 18. ·998.

EXERCISE 39

P. 300. 1. (a) 28; (b) 28. 2. (a) 137; (b) 136. 3. (a) 429; (b) 431.

EXERCISE 40

P. 301. 1. (a) 70°; (b) 4. 2. 50–54.

EXERCISE 42

P. 312. 1. $3x^2$; 2; 5; 7; $6x$; $70x^6$; $10x - 4$; $14x + 8$; $9x^2 - 4x + 1$.

2. 4; $- 6$; $32t$; $10 + 32t$; $80 - 32t$.

EXERCISE 43

P. 313. 1. $\frac{1}{2}x^2 + c$; $\frac{1}{4}x^4 + c$; $\frac{5}{3}x^3 + c$; $\frac{3}{2}x^2 + 4x + c$; $5x - x^2 + c$; $\frac{2}{3}x^3 + x^2 + x + c$; $3x - 2x^2 - x^3 + x^4 + c$.

2. $t^2 + c$; $\frac{3}{2}t^2 + 2t + c$; $4t - \frac{1}{4}t^4 + c$; $t^3 + t^2 - 7 + c$; $4t - 4t^2 + 2t^3 + c$.

INDEX